EXPERIENTIA SUPPLEMENTUM 22

Arthur Linder

Contributions to Applied Statistics

Dedicated to Professor Arthur Linder

Edited by Walter Joh. Ziegler

1976 Birkhäuser Verlag, Basel und Stuttgart

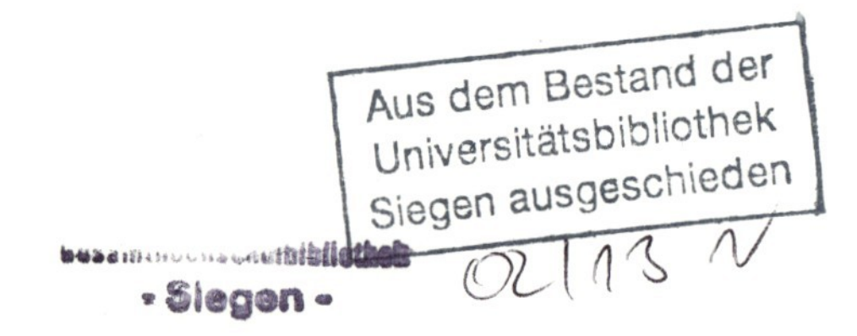

CIP-Kurztitelaufnahme der Deutschen Bibliothek

Contributions to applied statistics:
dedicated to Professor Arthur Linder / ed. by Walter Joh. Ziegler.
(Experientia: Suppl.; Bd. 22).
NE: Ziegler, Walter Joh. [Hrsg.]; Linder, Arthur: Festschrift.

ISBN 3-7643-0721-8

Foreword

In spite of all emphasis on objectivity, a scientific discipline needs for its development outstanding research workers with exceptional personal qualities. A scientist with these qualities is honoured here, on the occasion of his 70th birthday, by his pupils, friends and colleagues: Arthur Linder is respected as a researcher, advisor and teacher in the field of Applied Statistics.

Professor Dr. Dr. h. c. A. Linder has demonstrated the utility of statistical methods in science and technology; these are recognised today as being essential for an objective evaluation of data. His contacts with Sir R. A. Fisher, C. I. Bliss, G. M. Cox, W. G. Cochran, P. C. Mahalanobis and many other experts, have helped him to achieve his almost legendary authority on sampling and on the evaluation of statistical data.

Linder has won many friends and adherents to the cause of Biometry, Econometry and Technometry, firstly through his teaching activity at the universities of Berne, Geneva and Lausanne and the Swiss Federal Institute of Technology in Zurich, and as visiting lecturer at the Indian Statistical Institute, the International Statistical Education Centre in Calcutta, the University of North Carolina and the University of Natal in Durban, secondly through his contributions to the literature as author of the monographs 'Statistische Methoden' and 'Planen und Auswerten von Versuchen', as editor of METRIKA and co-editor of other journals, and thirdly through his active collaboration in the founding of The Biometric Society, to the presidency of which he succeeded R. A. Fisher in 1950–1951, and as inaugurator of the Austro-Swiss Region, over which he presided in 1964–1965.

In recognition of his outstanding achievements he has been elected Fellow of the American Statistical Association, Membre d'honneur de la Société Adolphy Quételet de Bruxelles, Member of the Review Committee of the National Sample Survey of India, Honorary Member of the Study Group for Operations Research of the Committee on Economical Production in Frankfurt, Honorary Fellow of the Royal Statistical Society of Great Britain, Fellow of the Institute of Mathematical Statistics, USA, and Honorary Member of the German Society for Operations Research.

The present volume is a modest attempt to reflect the multidisciplinary activities of A. Linder. The contributions convey an impression of how extensive and varied the field of applied statistics has become. Of course it was clear in compiling this volume that this impression would necessarily be incomplete. On the other hand it is highly gratifying to see how much original research work has been submitted by the various authors in response to the invitation to contribute. The enthusiasm of all those who have helped to compile this work is in itself a worthy tribute to Professor Linder as a friend, colleague and teacher. It is the editor's hope that these papers will stimulate critical and fruitful discussion.

The contributors were left entirely free in their choice of the subjects, and, so far as the actual presentation of their articles was concerned, they were merely requested to give generally comprehensive introductions or summaries in order to create some mutual understanding between structural and experimental scientists.

Unfortunately the number of invited papers had to be limited; on the other hand, a number of notable research workers were unable to contribute papers for various reasons, mainly because of the short deadline for the printing of this volume. Among those to whom thanks are due for their good intentions and their kind interest in the success of this volume are C. Bacher, E. Batschelet, R. K. Bauer, E. P. Billeter, C. I. Bliss, R. C. Bose, K. A. Brownlee, D. R. Cox, O. L. Davies, B. Day-Mauss, A. Hald, M. J. R. Healy, A. Jensen, H. Kellerer, M. G. Kendall, A. W. Kimball, S. Koller, R. Lang, F. E. Linder, D. Mainland, E. Olbrich, E. Pearson, R. Sailer, L. Schmetterer, H. Strecker, A. Vessereau, E. Walter, E. Weber, W. Winkler, F. Yates, and the authors of a special number of 'Statistik und Volkswirtschaft'.

In the name of the Council and the Advisory Committee of the Austro-Swiss Region of The Biometric Society, I wish to express many thanks to the contributors, to Mr. C. Einsele of Birkhäuser Verlag for his understanding and collaboration in the realization of this work and to those whose financial support has kept down the price of this volume.

Walter Joh. Ziegler

Contents

4. Testing

5. Relationships

6. Miscellaneous

1. General Outlooks

J. O. Irwin

J. O. Irwin

Some Aspects of the History of Biometric Method in the Twentieth Century with a Special Reference to Prof. Linder's Work

1. Introduction

It is an honour and a great pleasure to take part in the tribute to Professor Linder on his 70th birthday. I have long been interested in the history of our subject. My first paper on this topic was published in 1935, and I returned to it again, to some extent, in my presidental address to the Royal Statistical Society. In between these two dates, some fifteen years ago, I gave a talk at Pittsburgh to a joint meeting of several American societies. It was entitled 'Biometric Method, Past, Present and Future'. My object was to attempt a definition of biometric method, to show how it arose from developments, which go back as far as the sixteenth century; but that the main progress was made in the late nineteenth and in the present century. This subject is very pertinent to Professor Linder's work, for he is the 'Doyen' of Swiss biometricians and applied statisticians.

I do not now propose to cover all the ground previously covered, but only to call attention to some of the more important points. Where I felt that I could not improve upon what I said then, I have not hesitated to quote the 1959 paper.

2. Meaning of Biometry

First, what is biometry? We should all agree that it is something to do with *life* and something to do with *measurement*. As the word is now used, there is a third notion involved, that of *interpretation*. How widely the terms 'life' and 'measurement' are to be taken is open to discussion.

The term 'life' for this purpose may be extended to include any aggregate the individuals of which may vary and exhibit a certain (at least apparent) spontaneity of behaviour. According to A. N. Whitehead—whose later years were spent in the endeavour to build up a philosophical system adequate to describe the facts (aesthetic, cosmological and sociological) of the universe as it presents itself to modern man—the ultimate realities are 'occasions of experience'. Such an occasion of experience in *his sense of the term* may have nothing to do with living material. It is of the essence of each 'occasion of experience' that it is in part 'determined' and in part 'self creative'. On this view the ultimate realities possess a stochastic element. It would follow that biometric method need not be limited to any one field of subject matter. On this view one might, I think, say, that there is no distinction between Biometry and Applied Statistics.

The term 'measurement' can for this purpose, also be widely interpreted. It includes physical measurement in the strictest possible sense; it includes the recording of qualities, and also the region intermediate between the two. Perhaps, most important of all, it includes *counting*. One of the tasks of biometric method is to find the appropriate technique for dealing with each kind of measurement.

The *object* of biometric method is to make general statements about the properties of groups from measurements made on individuals belonging to them. This involves estimation, inference and interpretation. The function of mathematical statistics is to help to make this possible. Without mathematical statistics the subject could not exist. As we know it, it could not have been developed. Nevertheless I believe that mathematical statistics are necessary but not in general sufficient for the purpose.

3. Illustrations from Prof. Linder's Work

Professor Linder's work illustrates all these points very well.

Thus between 1931 and 1947 we find that most of his papers dealt with demography, with a very good balance between theory and its application. Here a number of the papers seem to be concerned with getting across to non-mathematicians essentially mathematical ideas—which he does very clearly. Good examples are gross and net reproduction rates.

During this period there is also one paper (1934) on body build and athletic performance while in 1946 and 1947 we find two overtly biometrical papers on *bee-research*.

In 1945 he published his book '*Statistische Methoden für Naturwissenschafter, Mediziner und Ingenieure*'. This contains all the essential statistical theory, well illustrated by examples, that would have been familiar to well qualified statisticians who had started out in the twenties and thirties (and I suspect *some theory* that would *not* have been familiar to them). Analysis of variance, discriminatory analysis, and generalised distance have all been included.

By this time Professor Linder had turned his attention to applications involving very diverse subject matter. For instance we find in the book examples from the fields of industrial manufacture and quality control, from medicine, biology and anthropometry, from agriculture and forestry, in addition to one or two from physics.

His papers from 1947 onwards continue to show this interest in applications in a large number of diverse fields, while 1953 was noteworthy for the appearance of his book on the design and analysis of experiments in the fields of natural science, medicine and enginneering. *(Planen und Auswertung von Versuchen. Eine Einführung für Naturwissenschafter, Mediziner und Ingenieure.)* We can observe his growing interest in biological assay and in sampling surveys. In the latter field he was a member of a Committee appointed by President Nehru to supervise the Indian National Sample Survey.

I introduced Section 3. at this point—though, historically, it is an anticipation—to show that all Linder's work could well be regarded as the development and application of biometric method. We now have to enquire how it came to have its truly tremenduous development in the twentieth century.

4. Two Main Streams of Derivation

In primitive man observation preceded conscious theorising. He looked out on the world and noticed the change in the seasons. As soon as primitive man or primitive woman said 'When we sow seed in the spring we get a crop in the autumn' an induction had been made and a hypothesis formed, one which could be tested by observation. Thus observation and theory have been developed together from the beginning. And so, as we might expect, there are, in modern biometric method, two main streams to the source of development. One goes back to the early workers at the beginning of the seventeenth century, who first thought of taking numerical measurements, recording qualities and then aggregating the results. The other goes back to the seventeenth and eighteenth century writers on the theory of probability and to Gauss in the early nineteenth century.

The former line of descent leads to Quetelet and Galton in the nineteenth century, nor must Florence Nightingale be forgotten. The two streams unite in Karl Pearson and his school, Edgeworth, and R. A. Fisher.

5. Development of the Observational Side of the Subject

It seems natural do deal first with the former source. If we use the most general sense of the term 'measurement' observation in biometry can be classified into the following three categories:

a) Measurements made in relation to a single individual.
b) Measurements or numerical statements relating to small groups of individuals.
c) Measurements or numerical statements relating to large groups of individuals, or even to populations as a whole.

We are not concerned with the first group as such. Yet we have to know how to measure a single individual, before we can deal with groups. This necessitated the invention of instruments of measurement. In medical statistics, for example, it seems that this started with Sanctorius of Padua (1561–1636) who had primitive instruments for measuring temperature and pulse-rate—and the development has continued until the present day.

Until about 1915 biometric method had concerned itself only with the third group. This group includes economic statistics in general and the whole subject of vital statistics and demography with its actuarial applications. It also includes, as we shall shortly see, nearly all the work of Karl Pearson and his school on biology and eugenics.

From the second group twentieth century biometrics has developed. This group is distinguished by the opportunities it provides for controlled experiment. Modern knowledge of the principles of experimental design was first applied in agricultural science, but has been used more recently in ever widening fields. Instances are the study and control of the physical and chemical processes of industrial production, problems of bacteriology and virology, psychological study of cognitive affective and conative processes or of fatigue, physiological studies such as the effect on animals of carcinogens, or of radiation, and psycho-physiological studies such as the effect of hot climates. Among the most noteworthy of the more recent fields of application are clinical trials and biological assay.

6. Development of Theory

It is clearly not possible in a relatively short paper to trace the development of probability theory through the late seventeenth, eighteenth and early nineteenth centuries. However, it is a fact that many of the results believed to be modern are to be found in the work of the early writers. For example, when—about 1925—I derived a general form for the sampling distribution of the mean by a Fourier inversion of what is now called the characteristic function, I thought I had found something new. But the result, for the rectangular distribution at any rate, was discovered by Lagrange. However the earliest writers confined their discussion to games of chance; the most noteworthy contributions to applied mathematics of the greatest figures of the eighteenth century such as Lagrange and Laplace were in other fields, mainly that of dynamics. De Moivre laid the foundations of actuarial method by his work on life annuities and Gauss developed the theory of errors of observation. But the biological field, as usually understood, was hardly explored at all.

We have to take a leap in time! The modern age of biometrical method started with Karl Pearson, born in 1857. Pearson started as a Cambridge mathematician. He was appointed to the Chair of 'Applied Mathematics' at University College London in 1884. His youthful interest in medieval literature, in law and socialism, lie outside the scope of this paper. The 'Grammar of Science' which was largely written in 1880 is, however relevant. One of the best accounts extant of what Science really does, it took the view that scientific laws are descriptions in 'conceptual shorthand' of data derived from our sense perceptions. With the nature of the realities behind the sense data Pearson refused to concern himself. He felt, perhaps unconsciously, that entanglement in metaphysical questions would have interfered with his power to develop, as he wished, the subject which he did in fact develop with immense drive.

Bio-statistical techniques can be divided into those which are exploratory and those which are aimed at testing theoretical models. Pearson *was* interested in scientific theories; sometimes, as in his famous controversy over Mendelism with Bateson, he attacked them violently. But one feels that the predominant contribution to biometry of Pearson and his school was in the field of exploratory techniques. The techniques of correlation and regression which he developed theoretically were of this kind. He and his followers felt that by applying them to sufficiently large bodies of data and studying the results, we might learn to understand the nature of the processes going on. The occasions when they tested particular hypotheses or 'models' were relatively rare.

Pearson's system of frequency curves was based on an extremely general hypothesis, that 'contributory cause groups' were at work, which would lead to hypergeometric distributions. This led him to the problem of representing such distributions by continuous frequency curves—which was solved by the use of the well known differential equation, expressing the parameters as function of the first four moments. His main use of them was exploratory to find empirical formulae for observed frequency distributions. In fact they have also frequently been found valuable in getting an idea of what a theoretical distribution is like when we can find its moments. Student (W. S. Gosset) discovered the exact sampling distribution of the variance from a normal universe and the 't' test by this method; in this case the result was later shown by Fisher to be exact. Exploratory techniques will always be essential to biometry. Perhaps the best example, among techniques currently used is that of 'factor analysis'. The viewpoint of the *Grammar of Science* accounts for the subsequent emphasis on exploratory techniques.

The year 1890 was also important to Pearson for another reason. Weldon was appointed to the chair of Zoology at University College London; this stimulated Pearson's interest in biology and it was probably through Weldon that he came to meet Francis Galton. This was the turning point in Pearson's life. Galton was one of the greatest scientific figures of the nineteenth century and the first in England to see the possibilities of statistical method applied to biology, especially to heredity, but also to medicine, anthropometry and psychology. He inspired Pearson with a lasting enthusiasm. Galton was not a mathematician and Pearson supplied the necessary specialised knowledge. Starting in the nineties Pearson published his 'Mathematical Contributions to the Theory of Evolution' (1893–1900) in which he developed the theory of correlation and regression and the Pearsonian system of frequency curves. These appeared in the *Philosophical Transactions of the Royal Society*, and after some years resulted in the controversy with Bateson over Mendelism. Resulting difficulties with the Royal Society over publication led to the foundation of *Biometrika* in 1901.

In the early years of the present century Pearson was still Professor of Applied Mathematics at University College, but he was running a Eugenics Laboratory and a Biometric Laboratory at the same time. He continued to have the most amazing energy. In 1911 Galton died and left money for the endowment of a Chair of 'National Eugenics'. Pearson then resigned the Chair of Applied Mathematics and became the first Galton Professor. He called his department the 'Department of Applied Statistics and Eugenics' and ran all these activities together: Statistical Theory and Practice *and* Biometry including Anthropometry and Eugenics. Ultimately he completed fifty years as a Professor at University College London. In 1957, the centenary of his birth, a Memorial Lecture was delivered by J. B. S. Haldane. The following passage from Haldane's tribute seems to me well balanced and just:

"I believe that his theory of heredity was incorrect in some fundamental respects. So was Columbus' theory of geography. He set out for China and discovered America. But he is not regarded as a failure for this reason. When I turn to Pearson's great series of papers on the mathematical theory of evolution, published in the last years of the nineteenth century, I find that the theories of evolution now most generally accepted are very far from his own. *But* I find that in the search for a self-consistent theory of evolution he devised methods which are not only indispensible in any discussion of evolution. They are essential to every serious application of statistics to any problem whatever. If for example I wish to describe the distribution of British incomes, the response of different individuals to a drug, or the results of testing materials used in engineering, I must start from the foundations laid in his memoir "Skew variation in homogeneous material". After sixty-three years I shall certainly take some short cuts through the jungle of his formulae, some of which he himself made in later years. Very few ships to-day follow Columbus' course across the Atlantic!"

"Let me put the matter another way. Anyone reading the controversy between Pearson and Weldon on one side, and Bateson and his collegues on the other, which reached its culmination about fifty to fifty-five years ago might have said 'I do not know who is right, but it is certain that at least one side is wrong'. In fact both were right in essentials. The general theory of Mendelism is, I believe, correct in a broad way. But we can now see, that if Mendelism were completely correct, natural selection, as Pearson understood it, would not occur. For the frequency of one gene could never increase at the expense of another, except by chance, or as we now put it, sampling errors. It is just the difference between observed results and theoretical expectations, to which Pearson rightly drew attention, which gives Mendelian genetics their evolutionary importance."

Edgeworth was slightly older than Pearson. He was for many years Professor of Economics at Oxford. While Pearson's work and Pearson's writing were characterized by definiteness and sharp clarity of outline (one might sometimes think he was wrong but one was never in doubt as to what he meant),

Edgeworth was courtly, urbane and diffident. His writing had a certain obscurity and there were frequent digressions with learned allusions and classical quotations. Perhaps that is why his statistical work, which was of firstclass originality and importance, is not better known than it is. He saw deeply into the subject. He discovered the general expression for frequency distributions in terms of the successive derivatives of the normal frequency function and used it to considerable advantage; and one can see foreshadowed in his work both maximum likelihood and analysis of variance.

A quotation from A. L. Bowley's account of his work illustrates his manner. Most of this is in Edgeworth's own words:

"In 1908 it is again shown 'how widely the subjective element enters into the calculus of probabilities' for the whole must in the end be related to credibility; similarly utility must be ultimately relative to the feeling of happiness, if it is to have an intelligible meaning. In both economics and statistics classical writers seem to have over-estimated the precision of their statements owing to 'undue confidence in untried methods of deduction'. There is even a similarlity between the particular instruments in earch department which have thus proved treacherous—here the postulate of complete independence between events, there the postulate of perfect competition between persons. There is common to both studies a certain speculative or dialectical character which recalls the ancient philosophies. 'In wandering mazes lost' too often both pursue inquiries which seem to practical intellects interminable and uninteresting. These characteristics, it is to be feared, may seem to attach to inquiries, like those pursued in an earlier portion of this paper, into the *a priori* probabilities of various measurements. In economics too, other people's mathematics are apt to resemble the

'... dark lantern of the spirit
which none see but those that bear it'."

Yule, Greenwood and 'Student' were all Pearson's pupils. Yule was the first. He ranks among the greatest of statisticians. He became lecturer in Statistics and a Fellow of St. John's College, Cambridge, where he remained until he died. He had learning, humour, strong common sense and originality. He put the theory of correlation and regression into a form in which it could be used by people of quite modest mathematical attainments. Much of this was incorporated in the famous textbook '*An Introduction to the Theory of Statistics*' in which the viewpoint is all his own. This the first textbook, still continues in its modern form 'Yule and Kendall' and for a general description of the field has never been surpassed. His originality is evidenced by his studies of literary vocabulary and his work with Willis on the distribution of species and by his study of Wolfer's sunspot numbers.

His treatment of the last two topics provide (apart from the work of McKendrick which I have described elsewhere) the earliest examples of stochastic process methods in the literature. Yule was well aware of the necessity all statisticians have, at times, to use exploratory methods with data which have no element of randomisation. His remarks on this should not be forgotten by any of us:

"The unhappy statistician has to try to disentangle the effect from the ravelled skein with which he is presented. No easy matter this, very often; and a matter demanding not merely a knowledge of method, but all the best qualities that an investigator can possess—strong common sense, caution, reasoning power and imagination. And when he has come to this conclusion the statistician must not forget his caution: he should not be dogmatic. 'You can prove anything by statistics" is a common gibe. Its contrary is more nearly true—you can never prove anything by statistics. The statistician is dealing with the most complex causes of multiple causation. He may show that the facts are in accordance with this hypothesis or that. But it is quite another thing to show how all other possible hypotheses are excluded, and the facts do not admit of any other interpretation than the particular one we have in mind."

W. S. Gosset, who for most of his life wrote under the pseudonym of 'Student' was a pupil of Pearson's, originally a chemist. He spent the whole of his working life with Guiness's of Dublin and ultimately came to be Chief Brewer at their London Brewery of Park Royal. The Statistical Department, which he started, must have been the earliest of industrial research departments. He early sensed the need for statistical method in industrial research, particularly the theory of small samples. He was not a professional mathematician; he obtained his theoretical results by simple algebra and extraordinary insight. The man who, by such simple methods, discovered the 't' test and the sampling distribution of the correlation coefficient in the *null* case deserves to be described—as he was described by Sir Ronald Fisher—as the Faraday of the subject.

A man of great modesty and outstanding charm 'Student' was never known to quarrel with anybody. He was on the friendliest terms with both Karl Pearson and R. A. Fisher, and thus he formed a bridge between the older and newer biometric schools.

The older schools had developed techniques which were mainly exploratory. Fisher developed the techniques necessary to make the subject useful in the field of scientific experiment. Further his work always emphasised the importance of building up relevant hypotheses, or as many people nowadays call them 'models' which may be tested by suitable techniques applied to the data of observation and experiment. In summary I should say that the importance of his work lies:

a) In the development of the theory of small samples (and therefore of samples of any size), in the course of which he obtained the *exact* forms of many sampling distributions.
b) In the systematic development of tests of significance.
c) The development of the theory of estimation, showing what can be done independently of probabilities *a priori*.
d) In laying down the principles of experimental design and providing us with a technique for analysing the results.

e) In the development of much of the technique of multivariate analysis and discriminant function analysis.

f) In the reconciliation of the Mendelian and biometric views of heredity by working out the consequences of particulate inheritance when applied to populations at large and in the development of statistical methods appropriate to genetics.

g) In much work on statistical inference, besides that involved in b) and c) above.

Sir Ronald Fisher was one of the few men whose fame became legendary in his own lifetime. Towards the end of the 1914–1918 war it seemed to Sir John Russell, who had become director of the Rothamsted Experimental Station, that they had a tremendous lot of figures dealing with about eighty years experiments, and perhaps a mathematician might be able to make something of them. So it came about that in 1919 Fisher went to Rothamsted, and started out, just with a table and a calculating machine, to see what he could make of the Rothamsted figures. In about ten years he had built up a statistical department, which was already becoming well known in a great many parts of the world.

How did he do this? Well, he was a mathematical genius and had great biological insight. He really understood the biologist's point of view and was tremendously quick in the uptake; also able very rapidly to switch an extreme degree of concentration from one subject to another—apparently quite effortlessly.

Now biologists, at any rate those of the calibre who were at Rothamsted in the twenties, are very intelligent people. They knew what they were doing and what they wanted to get at. They soon found, after talking with Fisher, that he understood their problems. When Fisher said to them 'Well if you go and do this or that sort of calculation with your results, you will get what you want', they did not in the least know how these particular techniques could be justified—but they went away and tried. They soon found they got what they wanted.

Whereever one goes in the world now, one finds experiments being designed and analysed by methods which are essentially Fisher's. I wonder sometimes how many of the statistical departments and institutes of the world would be making the progress they are making, were it not for the work of Fisher.

Fisher started as a Cambridge mathematician. To Cambridge he returned as Professor of Genetics. A few years after he retired from that chair, he went to live in Adelaide in South Australia, working in the late Dr. E. A. Cornish's department. There he died suddenly in 1962.

7. Conclusions

So far, apart from the references to Professor Linder himself, I have been dealing with the work of my seniors. I have said little about work done after 1960 and not much about that of my contemporaries.

However, I think this is the place to recall Neyman and E. S. Pearson's work on hypothesis testing. If it had done nothing more than introduce the notion of the *power* of a test, that alone would have made it of the greatest importance—and I believe it provided the basis for much theoretical work done subsequently, especially in America. It must have helped to inspire Abraham Wald.

The second world war was a great stimulus to the development of statistical and biometric method. On the theoretical side, this led, I think, through sequential analysis to decision functions and much of the work of the modern American school in statistical inference. This re-thinking of the problems of statistical inference is not now confined to America. The matter is still controversial; my own opinion, for what it is worth, is that the controversy will last as long as the subject itself.

The development during the contemporary period of stochastic process theory has been striking. Here we should not forget the earlier work of McKendrick, but I am thinking now, for example of the work of Feller, M. S. Bartlett, D. G. Kendall and their pupils. The epidemiological developments have been important, a good account is given in Norman Bayley's book. But whether we look at the theory of cosmic rays, demography, the growth of bacterial populations, population genetics, industrial renewal theory, queues or economic time series, we can find instances of the application of stochastic process theory, and signs that these are only forerunners of what is to come. My reference to the contemporary period would not be complete without mentioning the outstanding activity in clinical trials, biological assay and sampling surveys. To go into detail would double the length of this already long paper.

I should like to conclude on a personal note! My first academic appointment was in 1921. As far as I know, I am the only man living who served on Karl Pearson's and on R. A. Fisher's staffs. So I grew up with the subject and my life became intertwined with it. I first met Professor Linder in 1948 or 1949. We both participated in the I.S.I. and Biometric meetings at Berne (1949), in India (1951) and in Brazil (1955). In 1953 he persuaded me to give a lecture (the only one I have ever delivered in German) at the 'International Forum of Zurich'. On more than one of these occasions he was particularly kind to me. It therefore gives me great pleasure to have lived long enough to congratulate him on his seventieth birthday.

Author's address:
J. O. Irwin, Säntisstrasse 64, 8200 Schaffhausen.

Franz Ferschl

Franz Ferschl

Einige Bemerkungen zur Grundlagensituation in der Statistik

Zusammenfassung

Es wird versucht, den derzeitigen Stand der Grundlagendiskussion zum Inferenzproblem in der Statistik kurz zu charakterisieren. Die zu diesem Stand führende Entwicklung wird skizziert. Es zeigt sich, dass den Kern der Diskussion immer noch die Frage nach der Geltung des subjektiven Wahrscheinlichkeitsbegriffes bildet. Einige zentrale Argumente in dieser Auseinandersetzung werden herausgearbeitet. Abschliessend werden Hinweise auf Lösungsversuche gegeben, welche neuerdings von philosophischer Seite erfolgen und die zur Klärung in der Debatte zwischen subjektivistischem und objektivistischem Standpunkt beitragen könnten. Es zeigt sich dabei, dass vor allem noch an der Begründung eines objektiven Wahrscheinlichkeitsbegriffes gearbeitet werden muss.

Summary

It has been tried to characterize the actual state of discussion concerning the foundations of statistical inference. The lines leading to this stage of development are sketched. It appears, that the central point of discussion is still the validity of the concept of subjective probability. Some arguments, seemingly essential in this context, are presented. Finally references for attempts of solution are given, which are recently undertaken by some philosophers. It seems possible, that these attempts would give a contribution in making more transparent the controverse opinions of subjectivism and objectivism. Thereby it will be apparent, that there is to do a lot of work especially in clarifying the idea of objective probability.

In jüngster Zeit ist die Grundlagendiskussion unter den Statistikern wieder mehr aufgelebt. Das Interesse an diesen Fragen zeigt sich nicht nur an einer beträchtlichen Anzahl neuerer Publikationen, sondern auch an einem ganz offensichtlich vermehrten Engagement von Statistikern in Sachen Grundlagen der Statistik, genauer: des statistischen Inferenzproblems.

Ein Grund hiefür ist sicher das neuerliche Vordringen von «Bayesmethoden» in der Statistik. In den letzten anderthalb Jahrzehnten ist so ziemlich das ganze Modellarsenal der parametrischen Statistik, angefangen vom einfachen Mittelwerttest bis zur Schätzung simultaner Gleichungssysteme in der Ökonometrie von diesem Gesichtspunkt aus bearbeitet worden. Interessanterweise findet gerade der anwendungsorientierte Statistiker Gefallen an dieser Vorgangsweise. Belege hiefür, herausgegriffen aus einer schon sehr umfangreichen Literatur, seien Box und Tiao [1] für Technik und Naturwissenschaft, Tribus [7] für Ingenieur- und Betriebswissenschaften und Zellner [8] für die Ökonometrie. Dabei ist folgendes zu beachten: Innerhalb der statistischen Entscheidungstheorie, die heute auch von den meisten Statistikern als übergreifendes Theoriegebäude anerkannt wird, fanden die «Bayesmethoden» eine zufriedenstellende Einordnung durch ihre Charakterisierung als die zulässigen Strategien im statistischen Entscheidungsproblem. Diesem «gemässigten Bayesianismus» gegenüber nimmt jedoch die obengenannte Richtung einen radikalen Standpunkt ein: Man verzichtet darauf, Bewertungen und Entscheidungsfunktionen einzuführen und zieht sich auf die Verknüpfung von priori-Verteilungen und posteriori-Verteilungen mittels der Likelihood-Funktion zurück. Vorbereitet wurde diese Entwicklung offensichtlich durch das Werk von Jeffreys, insbesondere das Buch Jeffreys [4]. Dieser «strikte Bayesianismus» muss sich dann jedoch der alten Kritik stellen, welcher die Wahl von priori-Verteilungen, die dann meist auf einem subjektiven Wahrscheinlichkeitsbegriff beruhen, suspekt erscheint.

De Finetti hielt beim 39. Kongress des Internationalen Statistischen Instituts ein vielbeachtetes Referat: «Bayesianism: Its unifying Role for both the Foundations and the Applications of Statistics». Eindrucksvoll war die engagierte Parteinahme während der Diskussion dieses Referats, die das sonst recht friedlich agierende Volk der Statistiker in zwei deutlich erkennbare Lager spaltete. Sowohl das Referat als auch die Diskussion lieferten klärende Gesichtspunkte; klärend insofern, als daran anschliessend der Kern der heutigen Auseinandersetzung ziemlich scharf herausgearbeitet werden kann.

Beginnen wir zunächst mit den Gesichtspunkten, welche die verschiedenen Positionen in ihren extremen Ausprägungen charakterisieren. Hier wäre als erstes der «Alleinvertretungsanspruch» zu nennen, der bei aller Vorsicht in den sonstigen Formulierungen in de Finetti [3], Seite 1, zum Ausdruck kommt: «Bayesian standpoint is noways one among many possible theories, but is an almost self-evident truth, simply and univocally relying on the indisputable coherence rules for probabilities». Untermauert wird dieser Standpunkt durch den Hinweis auf die – im wesentlichen gültige – Äquivalenz zwischen Bayes-Strategien und zulässigen Strategien im Entscheidungsproblem (die Relevanz dieses Arguments wurde in der Diskussion bestritten), aber auch durch Verweis auf den Umstand, dass subjektive Elemente bei den Modellkonstruktionen der klassischen, «objektivistischen» Verfahren nicht auszuschliessen seien. Das umstrittene Gegenargument hiezu: Die Wahl subjektiver priori-Verteilungen und die Wahl von Modellannahmen seien von verschiedener Qualität. Das hauptsächliche Unbehagen an den Bayes-Verfahren bildet aber offensichtlich die unbestreitbare Tatsache, dass bei gleichen, in einem bestimmten statistischen Experiment gewonnenen Beobachtungsdaten die Ergebnisse wegen der individuell möglichen Wahl der priori-Verteilung nicht eindeutig sind. Verschärft wird dieser Einwand, wenn man sich berechtigt fühlt, die Frage nach der «Korrektheit» oder «Richtigkeit» der priori-Verteilung zu stellen (siehe hiezu etwa Blyth [2]). Es zeigt sich schliesslich, dass die Kontroverse doch wieder auf die Gültigkeit eines subjektiven Wahrscheinlichkeitsbegriffes hinausläuft.

Es sei nun der Versuch gewagt, die entscheidenden Argumente für Gesichtspunkte zur Anwendbarkeit des subjektiven Wahrscheinlichkeitsbegriffes zusammenzustellen.

a) Wenn subjektive Wahrscheinlichkeiten an Individuen (genauer: an die konkreten Situationen von Individuen) geknüpft sind, also deren jeweiligen Wissensstand wiedergeben, dann ist die Frage nach der Korrektheit hinfällig; es bliebe allenfalls die Frage nach deren korrekter Gewinnung. Dies könnte man als das Liberalitätsprinzip des Subjektivismus bezeichnen.

b) Es gibt keine unbekannten (subjektiven) Wahrscheinlichkeiten. Die priori-Verteilungen werden als «Basisaussagen» in das statistische Problem eingeführt, jedes Individuum kennt seine subjektiven Wahrscheinlichkeiten[1]). Man ist zunächst geneigt, diese These ernstlich in Zweifel zu ziehen. Dazu aber folgende Überlegung. Sicher ist es nicht möglich, die obige Behauptung in dem Sinn zu verstehen, dass jedes Individuum in jeder Situation auf Anhieb einer bestimmten Aussage den zugehörigen subjektiven Wahrscheinlichkeitswert angeben kann. Jedoch welchen Sinn hätte dann die Behauptung der Bekanntheit subjektiver Wahrscheinlichkeiten? Es ist günstig, diese Frage nicht isoliert für subjektive Wahrscheinlichkeiten zu stellen, sondern die Vorgangsweise in ähnlich gelagerten Situationen zu betrachten, etwa beim Phänomen der «Tonhöhe». Im allgemeinen wird nicht bezweifelt, dass es

- die unmittelbar erlebte, subjektive Empfindung «Tonhöhe» gibt;
- sinnvoll ist, nach *Messverfahren* für subjektive Tonhöhen zu suchen.

Setzt man minimale messtheoretische Postulate als gegeben voraus, so gelingt tatsächlich die Konstruktion einer Skala für subjektive Tonhöhen. Ähnlich ist die Lage in der Theorie des kardinalen Nutzens. Man erkennt, dass die Bekanntheitsvoraussetzung für subjektive Wahrscheinlichkeiten dann berechtigt ist, wenn

[1]) Mit Recht wird darauf hingewiesen, dass hier in der *Praxis* die Gefahr der «Verfälschung» gegeben ist. Durch die Einfachheit der formalen Handhabung ist man leicht geneigt, sich auf «Gleichverteilungsannahmen», genauer: «Nichtinformative priori-Verteilungen» zurückzuziehen. (Zumindest wird man sich meist im Bereich der konjugierten Verteilungen bewegen; siehe hiezu etwa Box und Tiao [1], S. 25ff., Jeffreys [4], S. 101ff.)

- man die Existenz von Wahrscheinlichkeitsbewertungen im «Subjekt» überhaupt anerkennt;
- es eine vergleichsweise gute Messtheorie für diese Objekte gibt.

Letzteres wird man kaum bestreiten können. Neben vielen anderen Begründungsversuchen sei auf die elegante, kardinalen Nutzen und subjektive Wahrscheinlichkeiten simultan einführende Vorgangsweise in Pfanzagl [5], Seite 195ff., hingewiesen.

Dennoch bliebe wohl ein nicht auszuräumendes Unbehagen, wollte man die Alleingültigkeit des bayesianischen Standpunktes (siehe oben etwa das de-Finetti-Zitat) hehaupten: Soll der empirische Wissenschaftsbetrieb ganz auf Informationsverarbeitung, basierend nur auf individuellen und damit grundsätzlich der Kritik und Diskussion entrückten Basisaussagen reduziert bleiben? Stegmüller [6], Seite 224 f., charakterisiert diesen Anspruch als «reduktionistisches Programm» und vertritt den metatheoretischen Standpunkt, dass der strenge Anspruch reduktionistischer Programme bislang immer zum Scheitern verurteilt war.

Es erscheint also berechtigt, neben dem – durch seine Geschlossenheit faszinierenden – Gebäude der subjektiven Theorie auch an einem «objektiven» Wahrscheinlichkeitsbegriff festzuhalten. Was sollte eigentlich mit einem solchen objektiven Wahrscheinlichkeitsbegriff ausgedrückt werden? Jedenfalls dreht es sich darum, Sachverhalte zu konstituieren, die unabhängig vom Wissen einzelner existieren und über die man durch intersubjektiv definierbare Verfahren etwas herausbekommen kann, wie etwa über den Schmelzpunkt von Wolfram. In dieser Sehweise *gäbe* es also unbekannte Wahrscheinlichkeiten. Freilich stellte sich heraus, dass die Begründung des objektiven Wahrscheinlichkeitsbegriffs schwieriger ist als man ursprünglich angenommen hatte. Begründungsversuche, wie etwa der v. Misessche, scheiterten meist daran, dass man versuchte, diesen Begriff auf einfachere und grundlegendere zurückzuführen.

Als besonders belastend erwies sich die Forderung nach der Verifizierbarkeit (ebenso auch der Falsifizierbarkeit) der Wahrscheinlichkeitsaussage, die bis jetzt immer auf Zirkelschlüsse geführt hat. Hier dürften jedoch Vorschläge richtungweisend und erfolgversprechend sein, die besonders ausführlich in Stegmüller [6] beschrieben und analysiert werden. Als Ausgangspunkt dient wieder ein vergleichendes Argument: Auch in anderen Fällen ist eine an enge Kriterien gebundene operative Definition von intuitiv einleuchtenden Begriffen, etwa in der Physik, nicht ohne weiteres gelungen. Man denke etwa an das berühmte Beispiel des Dispositionsbegriffes «wasserlöslich». Wegen dieser Schwierigkeiten hat man jedoch diese Begriffe nicht fallengelassen, sondern hat versucht, sie *neu* und besser zu analysieren. Damit ist grundsätzlich der Weg für eine sinnvolle Weiterarbeit am Problem des objektiven Wahrscheinlichkeitsbegriffes gewiesen, und es kann nicht geleugnet werden, dass zumindest erfolgversprechende Ansätze in dieser Richtung vorliegen. Genannt sei hier etwa Poppers Propensity-Begriff, der Versuchsanordnungen zugeordnet wird, und darauf aufbauend ein zur Exponentialverteilung führendes Axiomensystem von Suppes (siehe hiezu Stegmüller [6], Seite 245ff.).

Nach wie vor wird man also die Grundlagensituation in der Statistik so charakterisieren können, dass zwei Sehweisen Anspruch auf Beachtung erheben können. Eine Zusammenfassung unter einem höheren Gesichtspunkt wird man wohl erst von einer zukünftigen Metatheorie über «Theorie und Praxis» – entsprechend dem heute aufscheinenden Gegensatzpaar objektive und subjektive Wahrscheinlichkeit (inklusive Entscheidungstheorie) – erwarten dürfen.

Es gilt also, der Empfehlung zu folgen, die Tukey (1973) anlässlich der Diskussion des oben genannten de-Finetti-Referates unter anderem gegeben hat[2]: «... let me suggest that the Bayesian's tolerance for individual choices of different a priori probabilities could well be extended to a wide-spread tolerance for different formalizations of inference, especially for different actual situations and problems.»

Literatur

[1] Box, G. E. P., und Tiao, G. C. (1973): Bayesian Inference in Statistical Analysis. Addison Wesley, Reading.

[2] Blyth, C. R. (1972): Subjective vs. Objective Methods in Statistics. The American Statistician *3*, Vol. 26.

[3] De Finetti, B. (1973): Bayesianism: Its Unifying Role for Both the Foundations and the Applications of Statistics. Invited Paper 16. 2. für die 39. Session des Internationalen Statistischen Instituts in Wien.

[4] Jeffreys, H. (1948): Theory of Probability, 2nd ed. Oxford.

[5] Pfanzagl, J. (1968): Theory of Measurement. Physica-Verlag, Würzburg.

[6] Stegmüller, W. (1973): Personelle und Statistische Wahrscheinlichkeit, 2. Halbband, Springer, Berlin.

[7] Tribus, M. (1969): Rational Descriptions, Decisions and Designs. Pergamon Press, Elmsford.

[8] Zellner, A. (1971): An Introduction to Bayesian Inference in Econometrics. Wiley, New York.

Adresse des Autors:
Prof. Dr. Franz Ferschl, Statistisches Institut an der Universität Wien, 1090 Wien, Rooseveltplatz 6. Privatadresse: 1170 Wien, Dornbacherstrasse 25.

[2]) Ausnahmsweise erlaube ich mir hier, aus den zum Zeitpunkt der Abfassung dieses Artikels noch unveröffentlichten Diskussionsbeiträgen am ISI-Kongress namentlich zu zitieren.

J. M. Hammersley

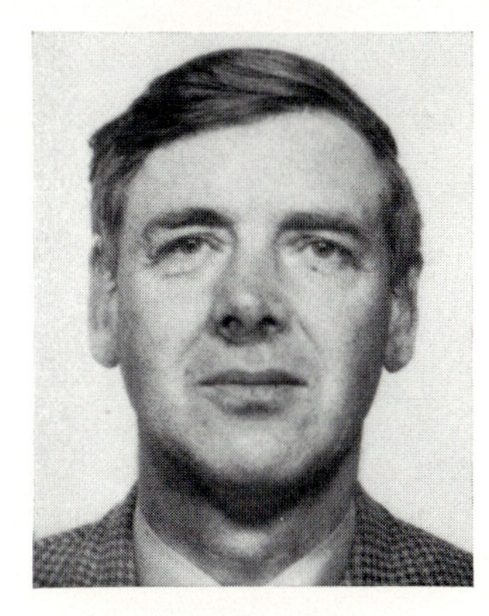

J. M. Hammersley

Some General Reflections on Statistical Practice

Summary

As one grows older, one gradually learns that statistical practice is more important than statistical theory. Applied statistics is a craft, not a doctrine.

In paying tribute to my friend and colleague, Professor Linder, I should like to set down a few personal reminiscences and to pass on from these to some general reflections on how a statistician's attitude towards his profession is liable to change as he grows older. I first had the pleasure of meeting Professor Linder when I was a young man: it was shortly after World War II and, with his encouragement, I was delivering a series of lecture in Zurich. It was my introduction to the international scene, the first time I had ever spoken abroad, and it was certainly very generous on his part to lend me his support because I had only just received my bachelor's degree a few weeks before and, moreover, had never had any formal training in the topics on which I was lecturing. But in those days statistics was not a subject which featured as a regular part of the normal university curriculum. To be sure, there were a few centres in the world, like University College in London, which did provide regular courses of instruction in statistics; yet the majority of practising statisticians hat not been trained in that way and had instead acquired their knowledge through a practical apprenticeship in some field of application. In my own case I had some experience of using statistics in certain military contexts during the war, and I shall always value my good fortune in having been just old enough to have belonged to that generation who embarked on their profession without any specialised academic preliminaries.

In these lectures at the ETH in Zurich I dwelt partly upon certain procedures for industrial quality control, especially those depending upon sequential sampling. These had only recently been developed in England and America and were then something of a novelty. An account of this part of the lectures was subsequently published [2] in Switzerland by the ETH. But I also included in the lectures some other (much more routine) material on the theory of significance tests and, looking back on the event, I can only smile and wonder at my motives for this.

One motive, naturally, must have been that it is much easier to lecture on theory than practice. Theory, especially when treated as a branch of pure mathematics, is cut and dried; the theorems follow from the axioms; one need only read up and regurgitate the material from books or published papers. By dressing the stuff up with a few numerical examples from practical experimentation, one can give it an aura of usefulness and cogency. A young lecturer, such as I was, will rarely probe the more sensitive areas involving the relationship and the validity to real life of either the axioms or their conclusions. However, I also recognise there must have been another motive too, the unspoken but nevertheless accepted motive of respectability; in an academic setting, topics appear respectable if bolstered by some fabricated theory. This belief and semblance run deep and are traditional, and have been embraced by statisticians much more distinguished than I can ever hope to be, and indeed by men of the very greatest eminence. How otherwise can one account for the fact that a man like Sir Ronald Fisher, whose statistical education was self-taught in the very practical and earthy milieu of Rothamsted, should have devoted one of his early papers [1] to the *mathematical* foundations of *theoretical* statistics? Or that an early major paper [4] of Jerzy Neyman should be *theory* 'based on the classical *theory* of probability'?

Looked at in retrospect, the achievement of Fisher and Neyman are monumental because they brought to scientific fruition the *practice* of handling data subject to random error, but not because they established *theories* which justified such procedure. Indeed, with the passage of time, the practical procedures have grown in popularity while the corresponding theories have wilted. By way of illustration let us look quite briefly at two familiar theories of significance testing—the Neyman-Pearson confidence theory and Fisher's fiducial theory.

Confidence theory is mathematically sound and consistent and rests on the ordinary axioms of probability, namely the axioms formulated by Kolmogorov. In its simplest form of testing a null hypothesis against an alternative it leads to measures of the so-called errors of the first and second kinds: if the null hypothesis is in fact true, we can calculate the probability that the test procedure will reject its truth; and if the null hypothesis is in fact false, we can calculate the probability that the test procedure will deny its falsity. But these two probabilities do *not* yield answers to the question which the everyday experimentalist asks the statistician. For the experimentalist wants to know which of his *own* two particular hypotheses he should accept. It is no answer to tell him that, *if* he knew the answer to this question, he could then correctly choose between two probabilities about the long-run reliability of the statistician's test procedure. Indeed if he knew the answer to his question he would simply not have bothered to consult the statistician. What the statistician has done is to reject the experimentalist's question, and to substitute for it a question of his own relating to the long-run reliability of his own performance, to which he has given a mathematically correct answer. In short, the confidence theorist has played a confidence trick and given the right answer to the wrong question.

Fisher's fiducial theory on the other hand attempts to answer the right question by attaching probabilities, so-called fiducial probabilities, to the two hypotheses. Unfortunately, Fisher never gave a satisfactory definition of what he meant by fiducial probability, nor what axioms this sort of probability should satisfy. I remember asking him in person once whether fiducial probability satisfied the Kolmogorov axioms. First, he tried to bounce me by demanding to know what the Kolmogorov axioms were; and then, when I specified these axioms to him in detail, he abruptly changed the subject and refused to consider my original question any further. That might be considered an inappropriate method of pursuing scientific discourse, but one should remember that science is pursued by human beings, and is therefore subject to human fallibilities. Fisher was a great genius, but by a quirk of his own character he found himself unable to admit he was in the wrong over matters on which he had thought deeply. It turns out that fiducial theory has led to various mathematical contradictions in the literature. It is not easy to pin down the source of these inconsistencies because of the intangible nature of the foundations and the lack of specific axioms for the theory, and Fisher, gradually driven onto the defensive by these mathematical contradictions, retreated to a position in which fiducial theory became hedged about by qualifications and restrictions (such as *sufficient* statistics, which constitute a pretty limited class of estimators). I believe it is fair to say that fiducial theory is either mathematically unsound or else so circumscribed as to be largely inapplicable to most practical situations. In short, the fiducial theorist has given the wrong answer to the right question.

Of course, there is no shortage of other theories of hypothesis testing; but equally all have their limitations. Decision theory, for example, works well enough in confined economic situations like the inspection procedures mentioned earlier in this article; but comes to grief when one tries to prescribe a loss function for errors in, say, estimates of the velocity of light. For, in a sense, all theories are doomed to failure because they aspire to too much. In attempting to generalise, they pretend to embrace circumstances which are too heterogeneous to be encompassed in a neat theoretical package. Bayesian arguments, measures of personal preference, and the like, while superficially germane because hypothesis testing is ultimately a matter of human decision, cannot command agreement when no agreement exists on so multifarious a subject as human judgment, prejudice, preference, utility, desire, intention or motive.

As I have grown older I have tended more and more to reject and retreat from the pretensions of theory and to regard the analysis of statistical data as an *ad hoc* procedure, relying on quite simple numerical calculations each tailored to the individual circumstances of the experiment in hand. Statistical consultation is not a profession, nor a dispensation of a priestly dogma, but rather the humble pursuit of a craft. It is all very well for the giants of statistics, like Fisher and Neyman, to propound theory and even to profess it obdurately in the fact of inconsistencies; for giants cast shadows which are larger than reality. But the humbler practitioner ought to stay closer to his material to make it his business to *understand* the data and to interpret it simply in its own particular setting, to keep close to the ground and to cultivate his garden. Statistics is not a doctrine in its own right, but an adjunct service facility to other scientists. That is what I said at greater length when Neyman invited me to deliver a paper [3] on behalf of statisticians on the occasion of the opening ceremonies of the new building (Evans Hall) for his Statistical Laboratory at Berkeley; and service, if one can achieve it, is no mean thing.

References

[1] Fisher, R. A. (1921): On the mathematical foundations of theoretical statistics. Philosophical Transactions of the Royal Society of London *A 222*, 309–368.

[2] Hammersley, J. M. (1948): An elementary introduction to some inspection procedures. Schweizerische Zeitschrift für Betriebswissenschaft *11*, 315–322.

[3] Hammersley, J. M. (1971): No matter, never mind!. Bulletin of the Institute of Mathematics and its Applications *7*, 358–364.

[4] Neyman, J. (1937): Outline of a theory of statistical estimation based on the classical theory of probability. Philosophical Transactions of the Royal Society of London *A 236*, 333–380.

Author's address:
J. M. Hammersley, University of Oxford.

Charles A. Bicking

Charles A. Bicking

Side Influences of Biometricians

Biometricians have contributed much to statistical methodology that is of interest to workers in other fields. It has been well worth while, for instance, for statisticians in the physical sciences and in industry to read the biometric literature and to develop close contacts with outstanding biometricians like the man whom this volume is to honor. In the chemical and pharmaceutical industries, in particular, the lead of biometricians has been followed since the days of 'Student'.

The pioneering work of Bliss, Fisher, Linder, Snedecor and others provided models for the early introduction of design of experiments in industry. For many years not only the best but also the sole texts available to industrial statisticians, such as 'Experimental Designs' by Cochrane and Cox, included examples from the biological (agricultural) areas only. Cross fertilization of other areas was aided by meetings of the Biometric Society, the Gordon Research Conferences, and other organizations. Non-biometricians also published in Biometrics and in Biometrika and some of the classical papers of leading statisticians have appeared in their pages.

In the processing and evaluation of many commercial substances biometric principles are involved and opportunities are provided for industrialists to collaborate with biologists and agricultural or medical scientists. In the manufacture of organic chemicals, for example, studies by industrial statisticians embraced a variety of problems, such as chemical treatment of pine trees to induce (or retard) resin flow (forestry science), examination of insecticide residues in milk, in beef fat, or in truck crops (agricultural science and husbandry), and irritant effect on human skin of treatments given textiles (medical science).

One of the most obvious needs for biometrics in industry is in the evaluation of the effects of toxic dust, fumes, sprays, or chemicals on the health of workers. This interest in toxic materials extends to the effects of ash, fumes, or effluent wastes on human beings and other living and even inanimate things in the surrounding environment. Sometimes the problem never reaches the stage at which biometrics is required. Take for example the proposal to substitute popcorn for excelsior in packing fragile products for shipping. The Department of Agriculture ruled that even if the popcorn were dyed black, a health hazard was still presented by the possible contamination of the packing and its later use for food.

Industrial medical and personnel departments, as well as production management, are interested in measurement of hazards during production. The problems of smog, of stream pollution, and of corrosion of structures are well known. Product development, market research, and sales departments join hands with the medical department in evaluation of products for possible health hazards.

Thus, although his experience may have been primarily in process quality control or in the design of research experiments, the applied statistician in industry may find himself called upon to be a part-time biometrician, working with all segments of the enterprise.

Some of the methodologies used in solution of these biometric problems came straight from experience in using the methods of statistical quality control as applied to industrial production problems (control charts, attribute sampling). However, the importance of precision of sampling and testing and of the planning and analysis of experiments led to the use of methodologies more often covered in the biometric literature than in the quality control literature (significance testing, analysis of variance, and regression analysis). With time, of course, statistical quality control has become more sophisticated, with the development of evolutionary operation, proliferation of experiment designs and of multiple regression analysis, and dynamic control of processes. Nevertheless, a good industrial statistician, though he probably doesn't consider himself to be a biometrician, still finds the biometrics field to be an important resource source.

The wide influence of biometricians has extended not only to the vocational interests of others, but also to a wide spectrum of avocational interests. Although they are usually very modest about applications of statistics to their avocations, there are many non-biometricians who have made use of statistics in wild-life studies. A good example is a recent paper by Joan Keen of the Hirst Research Center, General Electric Company ('A life distribution for great tits—parus major', *Statistica Neerlandica, 26*, 3, August 1972). It is all the more interesting because it is a paper honoring Dr. H. Hamaker, another well-known industrial statistician who has an abiding interest in all matters concerning the contryside.

In avocational pursuits, the statistical practioner may be excused, perhaps, for bending the serious approaches of professional biometricians to what must often appear to be trivial matters. Sometimes these efforts are more practically inclined than at others. Collaboration in a Wild Life Service study of the incubation period of loggerhead turtles is one of the more serious efforts in the author's experience.

Data were recorded for 57 loggerhead turtle nests on the day of the season on which the nest was made and the length of the incubation period (with the help of William P. Baldwin who made a professional report on much wider aspects of the study). It was desired to determine the effect on length of incubination of the date of nesting. First, a graphic determination was made of the fact that the observations on length of incubation were normally distributed (so that no transformation of the data was needed before statistical analysis). Next, a scatter plot was made (arithmetic scales) of the number of days into the season the nest was made (x), against length of incubation period in days (y). The plot gave a strong indication of curvilinearity (had the pattern not been clear, a quadrant test could have been applied to give a preliminary graphic feel for the relationship). The following equation was postulated:

$$y = a + b\,x + c\,x^2 .$$

The solution of the simultaneous equations by hand by the substitution method resulted in the following regression equation:

$$y = 11.85 - 0.5164\,x + 0.0070\,x^2 .$$

The variance table for this solution is:

Source	S. S.	D. F.	M. S.	F	Significance
Days (x, x^2)	367.80	2	183.90	32.1	Beyond 99.9%
Residual	309.67	54	5.73		

The error of estimate (95% level) of length of incubation from date of nesting is + and − 4.65 days.

Many amatuer natural scientists have an interest in bird-study, for example, studying the migrational and population characteristics of song birds.

It may be desired to estimate trends in or limits of biological phenomena, such as the arrival dates of birds at nesting sites, using observations taken over very few seasons. Range statistics would be very useful. The standard deviation calculated from the average range of a large number of samples of n values each and a factor that expresses the ratio between the average range and the standard deviation or the universe sampled has been recognized widely as a useful 'inefficient' statistic. The ability to calculate the standard deviation from a single sample, though 'inefficient', would be useful.

If the population range is represented by R', an estimate of the standard deviation is obtained from

$$\sigma = R'/6 . \tag{1}$$

Two relationships between R' and a single sample range will be tried:

$$R' = 6\,R/\sqrt{n} \tag{2}$$

and

$$R' = (1 + 1/\log n)\ R' . \tag{3}$$

From Equation (2)

$$\sigma = R/\sqrt{n} . \tag{4}$$

From Equation (3)

$$\sigma = (1 + 1/\log n)\ R/6 . \tag{5}$$

The following table compares values of σ from Equations (4) and (5) for small n:

Estimate of Standard Deviation.

n	$\sigma = R/\sqrt{n}$	$\sigma = a\ R/6$[1]
2	0.707 R	0.721 R
3	0.578 R	0.517 R
4	0.500 R	0.443 R
5	0.448 R	0.405 R
6	0.408 R	0.382 R
7	0.377 R	0.363 R
8	0.353 R	0.352 R
9	0.333 R	0.341 R
10	0.316 R	0.333 R

[1]) $a = (1 + 1/\log n)$.

For the important sample sizes between $n = 3$ and $n = 8$, Equation (4) gives the more conservative estimate and is selected for trial.

For an example of application of the estimator $\sigma = R/\sqrt{n}$, we take, over a ten-year period, the arrival dates for the catbird (Dumetella carolinensis) at Wilmington, Delaware. Arrival varied from April 24 to May 8. The average was May 1, with a range of 14 days. Limits which bracket the probable arrival dates are obtained by multiplying this range by two (2) times the value of $1/\sqrt{n}$ for $n = 10$.

$$\begin{aligned} \text{Limits} &= \text{Average} \pm 2 \cdot 1/\sqrt{n} \cdot R \\ &= \text{May 1} \quad \pm 2 \cdot 0.316 \cdot 14 \\ &= \text{May 1} \quad \pm 9 \text{ days} \\ &= \text{April 22 and May 10.} \end{aligned}$$

Records over 32 years of the arrival of the Catbird in a neighboring county to the north are reported by F. L. Burns in 'The Ornithology of Chester County, Pennsylvania'. The dates reported were from April 22 to May 9, averaging May 2.

Based on only two season's observations, the limits for arrival of the Acadian Flycatcher (Empidonax virescens) in Delaware were May 6 $\pm$ 11 days, or April 25 and May 17.

This method would be of particular value in the study of new localities, in areas in which significant changes have been made in geographical features (new irrigations, large storage reservoirs, etc.), of species that are extending their range, or of symbiotic conditions not previously observed. Populations may be estimated on the basis of census counts and changes detected even from one year to the next.

The methodology useful to 'week-end' scientists tend to be those of a graphic nature and requiring simple calculations. It has to be fun! This only corresponds in sophistication to the precision of the instruments of observation likely to be available. In the loggerhead turtle study use was made of regression analysis following plots of frequency distribution and scatter diagram (with the option of using a quadrant test). Graphic displays of experimental results (two-way tables, comparison of mean effects, interaction diagrams) are also useful as a satisfying if not completely rigourous method of analysis.

The following data were collected on the hardening time of plaster of paris:

Hardening Time (Minutes).

		Normal water addition	Normal water addition + 10%
Type A	Start hardening	9.0	10.3
	Finish hardening	13.3	16.0
Type T	Start hardening	14.0	16.3
	Finish hardening	20.0	21.3

Two-way tables are constructed, as follows:

	Type A	Type T	Totals
Normal water	22.3	34.0	56.3
Added water	26.3	37.6	63.9
Totals	48.6	71.6	

	Start harden	Finish harden	Totals
Normal water	23.0	33.3	56.3
Added water	26.6	37.3	63.9
Totals	49.6	70.6	

	Start harden	Finish harden	Totals
Type A	19.3	29.3	49.6
Type T	30.3	41.3	71.6
Totals	49.6	70.6	

This is an inadequate design because all the factors are fixed but there is no replication term for estimate of error. Nevertheless, we shall make a visual comparison of the size of mean effects.

Effect	Mean difference	D.F.	F	Interaction assessment			
				W · T	W · H	T · H	W · T · H
Water add (W)	7.6	1	Probably				1
			Insignificant				1
Type (T)	23.0	1	Probably	0.4	0.4	1	1
			Significant	0.4	0.4	1	1
Hardening (H)	21.0	1	Probably				1
			Significant				1
			Av.	0.4	0.4	1	1
W · T	0.4						
W · H	0.4						
T · H	1	1	Insignificant				
W · T · H	1	3					
Error	Not available						

The main effect differences come from the marginal totals of the two-way tables. The two-way interaction assessments come from subtracting successively the differences between pairs in rows and columns within the two-way tables [Example: $T \times H$; $(29.3 - 19.3) - (41.3 - 30.3) = 1$ and $(30.3 - 19.3) - (41.3 - 29.3) = 1$)]. For the three-way interaction a similar process is followed within the original data table.

Interaction diagrams may now be constructed for the second-order interactions. Obviously, for $W \cdot T$ and $W \cdot H$ the lines of the diagram will be parallel (no interaction); for $T \cdot H$, the line for Type A will be one scale unit steeper than the line for Type T but will not be highly noticeable in this case. If the interaction were strong, the lack of parallelism or crossing of the lines would be a very vivid illustration of the interaction.

We have covered a spectrum of industrial and avocational applications of statistics to problems that come under the definition of biometric problems. We suspect that the number of such opportunities for persons not recognizable as biometricians is very great. The reason that there is such a wide-spread influence of biometricians is the fact that the biometric literature is quite generally comprehensible to people with some statistical training. More important, perhaps, is the willingness of men like Professor Linder to help those in other fields. In many unsuspected ways we come to appreciate the influence of a methodology and of a man.

Author's address:
Charles A. Bicking, 2635 Heritage Farm Drive, Wilmington, Delaware 19808.

2. Sampling

Tore Dalenius

Tore Dalenius

Sample-Dependent Estimation in Survey Sampling

Summary

This paper considers a problem which typically confronts survey statisticians, viz. how to cope in a sample survey with uncertainty concerning a parameter $\overline{X}$ of crucial importance to the choice of estimator of a parameter $\overline{Y}$. The paper considers an approach—sample-dependent estimation—characterized by using the sample itself to resolve the problem.

1. Introduction

1.1 *The Problem Situation*

As a means of giving a concrete form to the problem of sample-dependent estimation—SDE for short—I shall briefly discuss two particular cases.

1.1.1 Case No. 1

The mean $\overline{Y}$ of a variate Y is to be estimated for a certain population by means of a sample survey. The statistician in charge of the survey knows the mean $\overline{X}$ of a variate X which he believes is strongly correlated with the variate Y. Consequently, he constructs a sample design according to which $\overline{Y}$ is to be estimated by the ratio estimator:

$$\hat{y} = \frac{\overline{y}}{\overline{x}} \overline{X} ,$$

where $E\,\overline{y} = \overline{Y}$ and $E\,\overline{x} = \overline{X}$.

Having collected the observations, the statistician computes the sample correlation coefficient r_{YX}. He finds to his surprise that r_{YX} is 'small', suggesting that $\overline{y}$ may be preferable to $\hat{y}$ for the estimation of $\overline{Y}$. This finding actualizes the question: should he nevertheless use $\hat{y}$ as specified by the sample design, or should he use $\bar{y}$?

1.1.2 Case No. 2

The second case concerns a classic sample survey; for a detailed account reference is given to Jessen et al. [8].

The survey was carried out in Greece for the purpose of investigating the electoral lists which were being prepared for a plebescite. Pairs of observations y_i, x_i ($i = 1, \ldots, n$) were collected for a sample of first-stage sampling units. Plotting these sample observations suggested a particular model for the relation between the Y- and the X-variate, viz. $Y = B X$, with the variance of Y being proportional to X^2.

Those in charge of the survey had two options for the estimation of the total $T(Y)$ for the Y-variate:

$$t_1(Y) = \sum g_i y_i , \tag{1}$$

where g_i reflects the scheme used for the selection of first-stage sampling units;

$$t_2(Y) = b\, T(X) , \tag{2}$$

where b is the estimator of B, and $T(X)$ is the known total for the X-variate.

On the basis of comparisons of estimates of the standard errors of $t_1(Y)$ and $t_2(Y)$, $T(Y)$ was in fact estimated by $t_2(Y)$.

1.2 *Purpose of the Paper*

The two cases discussed in the previous section actualize the following question: is it satisfactory to leave it to the sample to make the choice of estimator?

The problem involved is one which has as yet received comparatively little attention from survey statisticians[1]). While *practical* considerations tend to favor a positive answer to the question raised, the use of SDE is not explicitly supported by survey sampling theory as presented in today's textbooks.

It is the modest purpose of this paper to provide some additional support for a positive answer in situations, where a choice *has* to be made[2]).

2. Toward a Theory for SDE in Survey Sampling

2.1 *The Setting of the Discussion*

In what follows, I shall view the problem of developing a theory for SDE as a problem in the realm of sample *design*. In doing so, I shall tackle the problem by means of two different approaches, viz.

a) empirically, and
b) analytically.

1) See, however, Tomlin [10] for an exception to this observation.

2) As briefly discussed in Dalenius [3], p. 77, a *combination* of two given estimators may be preferable to either one of them.

These approaches—which complement each other—are the subjects of sections 2.2 and 2.3 respectively.

2.2 *A Monte-Carlo Study*

In Dalenius and Peiram [5], a small-scale Monte-Carlo study of the performance of SDE is presented. The study concerns the choice between the usual mean-per-unit estimator $\bar{y}$ and the ratio estimator $\hat{y}$. The crucial parameters for this choice are:

$$R_c = \frac{1}{2}\left[\sigma_X/\bar{X} : \sigma_Y/\bar{Y}\right]$$

and the population correlation coefficient $R_{YX} = R$. For $R > R_c$, $\hat{y}$ is to be preferred to $\bar{y}$; for $R \leq R_c$, $\hat{y}$ is to be preferred to $\bar{y}$.

20 samples of $n = 100$ pairs of observations were selected from bivariate normal populations for 55 combinations of R_c- and R-values. For each sample, estimates r_c of R_c and r of R were computed. If $r > r_c$, then $\hat{y}$ was chosen to estimate $\bar{Y}$; otherwise $\bar{y}$ was chosen. The following diagram illustrates the kind of results which were observed: the heavy line denotes the ideal 'operating characteristic curve' (corresponding to the choice of $\bar{y}$ for $R \leq R_c$ and the choice of $\hat{y}$ for $R > R_c$); the broken line corresponds to the actual outcomes.

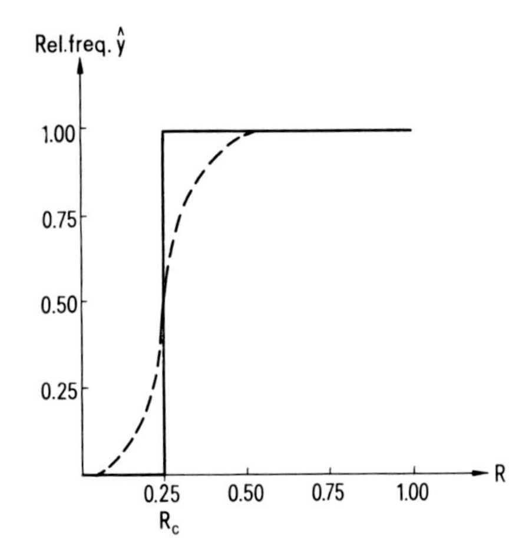

Fig. 1
Ideal vs. actual choice of estimator.

The findings of this Monte-Carlo study were summarized as follows in the reference cited above: 'The scope of the experiment is clearly restricted. It supports, however, cogently the conjecture that sample-dependent estimation should prove efficient, especially when dealing with large samples.'

2.3 *A Suggested Analytical Approach*

The intense debate in the last few years about the foundations of survey sampling has served to underline the importance of paying due attention to the *specific* features of survey sampling; what is 'optimum' in the context of sampling from a infinite population is not necessarily optimum when sampling from a finite population.

Considering SDE in the general context of statistical inference theory may throw some light on the specific problem of this paper. Therefore, in section 2.31, I shall present a brief review of SDE in that context. Next, in section 2.32, I shall return to the problem of SDE in survey sampling.

2.31 SDE in the context of statistical inference theory

It is interesting to note that on at least one occasion R. A. Fisher proposed the use of what amounts to SDE. In Fisher [6][1]) he examined a claim that—as a basis for estimating the mean error σ of a series of observations—σ_1 (based on the *absolute* deviations) was preferable to σ_2 (based on the *squared* deviations). Fisher examined σ_1 and σ_2, and showed that the choice should depend on the population observed: if the observations were from a normal population, σ_2 is the appropriate estimator, while if they were from, say, a double exponential population, σ_1 is the appropriate estimator.

Fisher closed the paper by discussing what to do in a case where it is known that the observations are from *either* a normal *or* a double exponential population. He suggested that the sample measure of kurtosis be calculated and went on to say:

> If this is near 3 the Mean Square Error will be required; if, on the other hand, it approaches 6, its value for the double exponential curve, it may be that σ_1 is a more suitable measure of dispersion.

In the half-century following Fisher's 1920 paper, the subject of sample-dependent inference seems to have received very little attention. In the last few years, however, the interest in the subject—often referred to as *adaptive inference*—has grown considerably. Three key references are Andrews et al. [1], Bancroft [2] and Hogg et al. [7]; these papers provide additional references of relevance in this context.

2.32 SDE and survey sampling theory

The success of the endeavors to build a theory for adaptive inference supports the viewpoint that a corresponding theory may be built for the case of survey sampling. I shall briefly discuss the format of such a theory; the discussion follows closely that in Dalenius and Fisz [4].

Consider a finite population characterized by an unknown parameter $\bar{Y}$ which is to be estimated by a sample survey. The sample design D to be used may be conceived of as:

$$D = S \times E .$$

[1]) My attention was drawn to this reference by Stigler [9].

Here S denotes that component of the design which deals with the procedures for getting observational access to the population (the construction of a hierarchy of sampling units, and the scheme for selecting a sample at each stage of this hierarchy). Likewise E denotes that component of the design which deals with the procedure(s) for estimating $\bar{Y}$, given the observations collected.

In a real-life situation, the options available to the survey statistician may look as follows:

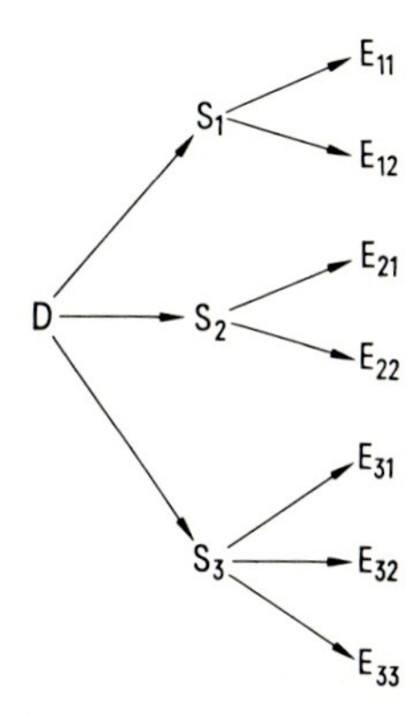

The design of a survey calls for first choosing S_i; this choice should clearly take into account the options with respect to the E-component! Next, it calls for choosing the estimator E_{ij}; this choice should be made to reflect the value of the unknown parameter $\bar{X}(S_i)$. More specifically, if $C[\bar{X}(S_i)]$ denotes the set of possible values of $\bar{X}(S_i)$, $C[\bar{X}(S_i)]$ may be partitioned in such a way $C_j[\cdot]$, $j = 1, \ldots, k$, that E_{ij} *is* the right estimator, if $\bar{X}(S_i) \varepsilon C_j[\cdot]$. SDE is equivalent to considering *stochastic* partitions of $C[\cdot]$, to be denoted by $C_j^*[\cdot]$, and using the estimator E_{ij} if $\bar{X}^*(\cdot) \varepsilon C_j^*[\cdot]$, where $\bar{X}^*(\cdot)$ denotes the estimate of $\bar{X}(\cdot)$.

References

[1] ANDREWS, D. F., BICKEL, P. J., HAMPEL, F. R., HUBER, P. J., ROGERS, W. H. and TUKEY, J. W. (1972): Robust Estimates of Location: Survey and Advances. Princeton University Press, Princeton, N. J.

[2] BANCROFT, T. A. (1972): Some recent advances in inference procedures using preliminary tests of significance, in: Statistical Papers in Honor of George W. Snedecor (ed. T. A. Bancroft). Iowa State University Press, Ames, Iowa.

[3] DALENIUS, T. (1957): Sampling in Sweden. Contributions to the Methods and Theories of Sample Survey Practice. Almqvist and Wiksell, Stockholm.

[4] DALENIUS, T. and FISZ, A. (1972): A note on sample-dependent estimation. Report No. 59 of the research project Errors in Surveys. Institute of Statistics, University of Stockholm.

[5] Dalenius, T. and Peiram, L. (1969): Sample-dependent estimation. Report No. 25 of the research project Errors in Surveys. Institute of Statistics, University of Stockholm.

[6] Fisher, R. A. (1920): A mathematical examination of the methods of determining the accuracy of an observation by the mean error, and by the mean square error. Monthly Notices of the Royal Astromomical Society, 758–770.

[7] Hogg, R. V., Uthoff, V. A., Randles, R. H. and Davenport, A. S. (1972): On the selection of the underlying distribution and adaptive estimation. Journal of the American Statistical Association, 567–600.

[8] Jessen, R. J., Blythe, R. H., Jr., Kempthorne, O. and Deming, W. E. (1947): On a population sample for Greece. Journal of the American Statistical Association, 357–384.

[9] Stigler, S. M. (1973): Laplace, Fisher, and the discovery of the concept of sufficiency. Technical Report No. 326, Department of Statistics, University of Wisconsin, Madison.

[10] Tomlin, P. H. (undated): An optimal ratio-regression-type estimator. Unpublished MS, Research Center for Measurement Methods, U.S. Bureau of the Census, Washington, D.C.

Author's address:
Tore Dalenius, University of Stockholm.

W. Edwards Deming

W. Edwards Deming

On Variances of Estimators of a Total Population Under Several Procedures of Sampling

1. Introduction

The aim here is to compare several simple plans of sampling that often appear to be equal, but which may give widely different degrees of precision when put into use. For example, it is well known that if we draw with equal probabilities and without replacement a sample of pre-determined size n from a frame of N sampling units (Plan I ahead), and if N be known and used in the estimator $X = N\,\bar{x}$, in the notation set forth in the next section, then X is an unbiased estimator of the total population A of the frame, and the Var X is given by Equation (5) ahead.

It is also well known that if the frame contains a proportion Q of blanks (sampling units that are not members of the universe), then the variance of an estimate of the total of some extensive characteristic of the frame increases out of proportion to Q, while the variance of the ratio of two characteristics suffers only from the diminished number of sampling units that come from the universe.

Not so well known is the effect of certain tempting procedures of selection in which the size n of the sample turns out to be a random variable. The purpose here is to examine and compare some of the alternatives.

One special case of importance is where one aim of the study is to estimate the total number N of sampling units in the frame. We first of all need some notation.

Notation:

P — probability before selection that any sampling unit in the frame will fall into the sample. In Plan I, P is the so-called sampling fraction. $Q = 1 - P$.

N — number of sampling units in the frame.

n — number of sampling units in the sample in Plan I and in Plan III.

$\hat{n}$ — number of sampling units in a particular sample in Plan II.

a_i the x-population of sampling unit No. i in the frame. a_i will be 0 if sampling unit No. i is not a member of the universe. a_i may also be 0 even if sampling unit No. i is a member of the universe.

$A = \sum^N a_i$ the total x-population in the frame.

$a = A/N$ the average x-population per sampling unit in the frame, including 0-values of a_i.

$\sigma^2 = \frac{1}{N} \sum^N (a_i - a)^2 = a^2(C_1^2 + Q)$ the overall variance between the a_i in the frame, including the 0-values of the a_i.

C_1 the coefficient of variation between the non-zero a_i.

$C = \sigma/a$ the coefficient of variation between the a_i in the frame, including the 0-values of a_i.

We first compare two plans, which we shall call Plan I and Plan II, for estimation of the total x-population of a frame: later, for estimation of a ratio. In both these plans the probability that a sampling unit will be selected into the sample will be P. In both plans we presume the existence of a frame, N known in some problems, not known in others.

2. Estimates of a Total Population

Plan I. n fixed at $n = NP$. N known. To select the sample, read out n unduplicated random numbers between 1 and N. This plan is sometimes called simple random sampling. Record the sample as $x_1, x_2, \ldots, x_n$, in order of selection. Compute

$$\bar{x} = \frac{1}{n}(x_1 + x_2 + \cdots + x_n)\,, \tag{1}$$

$$X = N\bar{x}\,. \tag{2}$$

Then

$$E\,X = A\,, \tag{3}$$

$$E\,\bar{x} = a\,. \tag{4}$$

That is, X is an unbiased estimator of A, and $\bar{x}$ is an unbiased estimator of a.

$$\operatorname{Var} X = N^2 \frac{N-n}{N-1} \frac{\sigma^2}{n} \doteq N^2 \left(\frac{1}{n} - \frac{1}{N}\right) \sigma^2\,, \tag{5}$$

$$\operatorname{Var} \bar{x} = \frac{N-n}{N-1} \frac{\sigma^2}{n} \doteq \left(\frac{1}{n} - \frac{1}{N}\right) \sigma^2\,. \tag{6}$$

For the rel-variances

$$C_{\hat{X}}^2 = C_{\hat{x}}^2 \doteq \left(\frac{1}{n} - \frac{1}{N}\right) C^2 . \tag{7}$$

All this is well known. The proofs are in any book on sampling.

Plan II. P fixed, N may be known or unknown; $\hat{n}$ a random variable. To select the sample, start with sampling unit No. 1. Accept it or reject it, depending on a side-play of random numbers. For example, if $P = .01$, read out a 2-digit random number between 01 and 00, all 2-digit random numbers to have equal probabilities. Let 01 accept the unit, 02 to 00 reject it. Then go to sampling unit No. 2; read out another random number between 01 and 00 with the same side-play and same rule. Go to No. 3, then to No. 4, and onward through the whole frame to N, always with the same side-play.

$$E\,\hat{n} = N\,P\,, \tag{8}$$

$$\operatorname{Var}\hat{n} = N\,P\,Q\,. \tag{9}$$

We here define X by Equation (15) ahead and note that

$$E\,X = A\,, \tag{10}$$

so we have again an unbiased estimator of A, but here

$$\operatorname{Var} X = \frac{N\,Q}{P}(\sigma^2 + a^2)$$

$$= \frac{N^2\,Q}{E\,\hat{n}}(\sigma^2 + a^2)\,, \tag{11}$$

$$C_{\hat{X}}^2 = \frac{Q}{E\,\hat{n}}(C^2 + 1)\,. \tag{12}$$

The proofs will appear in a minute.

Proof of the expected value and variance of X in Plan II:

$$\begin{cases} \delta_i = 1 & \text{if sampling unit No. } i \text{ of the frame falls into the sample,} \\ = 0 & \text{otherwise.} \end{cases}$$

We note that δ_i is a random variable and that

$$\delta_i^2 = \delta_i\,, \tag{13}$$

$$E\,\delta_i^2 = E\,\delta_i = P\,. \tag{14}$$

Define

$$X = \sum a_i \, \delta_i / P \,. \tag{15}$$

[Here and henceforth all sums will run from 1 to N unless marked otherwise.]

This is equivalent to

$$X = \frac{1}{P} x \,, \tag{15a}$$

where x is the total of the x-values in the sample. Then

$$E\,X = \frac{1}{P} \sum a_i \, E\, \delta_i = \sum a_i = A$$

which is Equation (10).

$$\begin{aligned}
\operatorname{Var} X &= \sum (X - E\,X)^2 \\
&= E(\sum a_i \, \delta_i/P - A)^2 \\
&= E(\sum a_i \, \delta_i/P - \sum a_i)^2 \\
&= E[\sum a_i(\delta_i/P - 1)]^2 \\
&= E \sum a_i^2(\delta_i/P - 1)^2 + E \sum_{j \neq i} a_i \, a_j(\delta_i/P - 1)\,(\delta_j/P - 1) \\
&= \sum a_i^2 \, E(\delta_i^2/P^2 - 2\,\delta_i/P + 1) + 0 \quad [\text{Because } \delta_i \text{ and } \delta_j \text{ are independent}] \\
&= \sum a_i^2(P/P^2 - 2\,P/P + 1) \\
&= \frac{Q}{P} \sum a_i^2 = \frac{Q}{P} \sum [(a_i - a) + a]^2 \\
&= \frac{N\,Q}{P} (\sigma^2 + a^2)
\end{aligned}$$

which is Equation (11)[1].

[1] I am indebted to my friend William N. Hurwitz, deceased, for this proof of Equation (11).

Remark 1. We pause to note that the difference in variances between Plans I and II may be alarming, or it may be inconsequential. To compare their variances, we set $E\,\hat{n}$ in Equation (11) equal to n in Equation (5) and write

$$\frac{\operatorname{Var}(\mathrm{II})}{\operatorname{Var}(\mathrm{I})} = \frac{\sigma^2 + a^2}{\sigma^2} = 1 + a^2/\sigma^2 \to 1 \quad \text{as} \quad a/\sigma \to 0\,. \tag{16}$$

This equation tells us that Plan II will always yield variance higher than Plan I will yield, and that the difference will be small only if a be small compared with σ. We shall return later to this comparison when we study the effect of blanks (0-values of a_i in the frame).

Remark 2. We note that for Plan II, $E\,x_i^2 = (1/N)\,\Sigma\,a_i^2 = \sigma^2 + a^2$ for any member i of the sample. Hence any x_i^2 in the sample is an unbiased estimator of $\sigma^2 + a^2$, and a sample of size $n = 1$ provides an estimate of $\operatorname{Var} X$ (noted privately by my friend and colleague the late William N. Hurwitz).

Remark 3. The appendix shows for illustration all the possible samples of $n = 1$ for $P = Q = 1/2$ that can be drawn from a frame of $N = 2$ sampling units, along with calculations and comparisons with some of the formulas just learned, and with some that will appear in section 5.

Plan III. Here, we separate out in advance the blanks, or attempt to do so. This plan has advantages and disadvantages. The required separation (screening) is sometimes costly. Plan III should be chosen only after careful computation of the expected variances and costs. An example and references appear later.

An example of Plan II. The problem is to estimate the total number N of fish that traverse a channel in a season. A shunt provides an alternate path, attracting into the shunt some average fraction P of the fish. It is a fairly simple matter throughout the season to count the fish that traverse the shunt, but not so easy to count the fish that traverse the channel. However, it is possible to count on a few selected days the fish that traverse the channel. Comparison of the counts of fish that traverse shunt and channel provides an estimate of P. The variance σ^2 between sampling units would be 0, as every sampling unit in the frame has the value 1. Then under the assumption that P is constant through the season, we could set $X = \hat{N} = \hat{n}/P$ for an estimate of N, where $\hat{n}$ is the number of fish that traversed the shunt during the season. Equation (12) would then give the conditional

$$\text{Rel-Var}\,\hat{N} = \frac{1 - P}{\hat{n}}\,. \tag{17}$$

Of course, the estimate $\hat{N} = \hat{n}/P$ would be no better than the estimate of P derived from the ancillary studies, but the estimate of the rel-variance of $\hat{N}$ derived from Equation (17) would be excellent if P be small.

Estimates could be made by direction of flow, upstream and downstream separately, and by big fish and little fish, if desired. The equation just written would give the conditional rel-variance of the estimate of any class of fish.

Another example. Any scheme for reduction of the probability of selection of sampling units that have some specified characteristic (such as certain items of low value) by use of random thinning digits or their equivalent should be examined carefully for the hazards of extra variance in the estimate of a total. One must weigh the simplicity and variance of Plan II against the lesser variance and possible extra costs of using Plan I.

A specific example of blanks may be described as follows. Suppose that the frame consists of $N = 3000000$ freight bills filed in numerical order in the office of a carrier of motor freight (possibly the inter-city hauls for one year). The management needs a sample of these shipments in order to study relations between revenues, rates, and costs as a function of weight, size, distance, and other characteristics of shipments. We suppose that the sample desired is 1 in 50 of the shipments that weigh 10000 lbs. or over, and 1 in 500 of those that weigh less than 10000 lbs. To make the selection, we list from the files on a pre-printed form 1 shipment in 50 (a systematic selection of every 50th shipment would serve the purpose); retain for the final sample every shipment listed that weighs 10000 lbs. or over, and select with probability 1 in 10 all other shipments. Suppose that the probability of 1 in 10 is achieved by preprinting the form with the symbol S on 1 line in 10, in a random pattern. Lines 1-11 on the form, when filled out, might appear as shown in Table 1.

Table 1

Line	Serial number	Weight (lbs.)	Remarks
1	CH 105474	2650	Not in sample
2	CH 105524	24450	In sample
3	CH 105574	220	Not in sample
4	CH 105624	175	Not in sample
5	CH 105674	800	Not in sample
6	CH 105724	720	Not in sample
7	CH 105774	15500	In sample
8 S	CH 105824	2750	In sample
9	CH 105874	120	Not in sample
10	CH 105924	13300	In sample
11 S	CH 105974	700	In sample
etc.			

The procedure of preprinting a form is tempting for its simplicity. But let us look at the variance of the estimate of (e.g.) the total revenue from shipments under 500 pounds. Let x be the aggregate revenue in the sample from these shipments. Then

$$X = 500\,x$$

will be an unbiased estimate of the revenue in the frame from shipments under 500 pounds. Unfortunately, Var X is afflicted with the term a^2 in Equation (11). The symbol a is the average revenue per shipment, for shipments of all weights, and $P = 1/500$. In practice, σ/a may be anywhere from .25 to .60. The term a^2 thus adds substantially to the variance of X.

A way out is to stratify into two strata the preliminary sample consisting of every 50th shipment, the two strata being (1) 10000 lbs. or over, and (2) under 10000 lbs. The 11 freight bills in Table 1 would now appear in two columns, as in Table 2. The symbol S in Table 1 is no longer needed: we take into the final sample every shipment listed under Stratum 1, and a selection of 1 from every consecutive 10 of the shipments listed under Stratum 2. We may form the estimate X as above, and the term a^2 in the variance will now disappear.

Stratification, serialization, and selection all require care, time, and supervision. Moreover, in practice, in the application to motor freight, there are 6 strata, not 2, with consequent enlargement either of errors or of care and supervision.

Table 2

Line No.	Serial number	Stratum 1 10000 lbs. or over	Stratum 2 under 10000 lbs.
1	CH 105474		2650
2	CH 105524	24450	
3	CH 105574		220
4	CH 105624		175
5	CH 105674		800
6	CH 105724		720
7	CH 105774	15500	
8	CH 105824		2750
9	CH 105874		120
10	CH 105924	13300	
11	CH 105974		700

When the record of shipments is on a tape, it is possible to stratify the shipments accurately in a number of strata and to select the sample from any stratum with a fixed proportion, thus eliminating the random character of the sizes of the samples. The extra cost is negligible if the stratification and selection be carried out along with other tabulations, all in one pass of the tape.

3. Estimates of a Ratio

A sampling unit has not only an x-value but a y-value. Thus, a sampling unit might be a household, b_i the number of people therein in the labor force, a_i the number of people in the household that are in the labor force and unemployed. Then

$$B = \sum b_i \tag{18}$$

is the total number of people in the labor force, and

$$A = \sum a_i \quad (19$$

is the total number of people in the labor force and unemployed. Put

$$a = \frac{A}{N} \tag{20}$$

[the average number of people per household in the labor force and unemployed]

as before, and

$$b = \frac{B}{N}\,. \tag{21}$$

[the average number of people per household]

Then

$$\varphi = \frac{A}{B} = \frac{a}{b} \tag{22}$$

is the overall proportion of people in the labor force unemployed. Suppose that we wish to estimate this proportion.

Plan I and Plan II both give estimates of A, B, and of $\varphi = A/B$. After seeing the possible losses in the use of Plan II for estimation of a total population, one may be astonished to learn that (so far as we carry our approximations to variances) Plan II gives for estimation of a ratio the same variance as Plan I, for a given size of sample. The proof will follow.

Plan I for a ratio. We first define the x- and y-variances between sampling units as

$$\left.\begin{aligned} \sigma_x^2 &= \frac{1}{N}\sum^{N}(a_i - a)^2 \\ \sigma_y^2 &= \frac{1}{N}\sum^{N}(b_i - b)^2 \end{aligned}\right\} \tag{23}$$

and the covariance

$$\sigma_{xy} = \frac{1}{N} \sum^{N} (a_i - a)(b_i - b) . \tag{24}$$

A sample drawn and processed by Plan I gives unbiased estimates of X and of Y by use of Equation (2). It gives also the ratio

$$f = \frac{X}{Y} = \frac{\bar{x}}{\bar{y}} \tag{25}$$

as the sample analog of $\varphi = A/B$. For the variance of f, we shall be satisfied with the usual Taylor approximation wherein

$$\text{Rel-Var} f = \text{Rel-Var} \frac{X}{Y} = \left(\frac{1}{n} - \frac{1}{N}\right)(C_x^2 + C_y^2 - 2\,\varrho\, C_x\, C_y) \tag{26}$$

which is satisfactory if n is big enough. Here

$$C_x = \frac{\sigma_x}{a} \tag{27}$$

[the coefficient of variation between all the a_i in the frame],

$$C_y = \frac{\sigma_y}{b} \tag{28}$$

[the coefficient of variation of all the b_i in the frame],

and

$$\varrho = \frac{\sigma_{xy}}{\sigma_x\, \sigma_y} \tag{29}$$

is the correlation between the N pairs of values a_i and b_i.

Plan II for a ratio. Again, $E\,X = A$ by Equation (10). Also, $E\,Y = B$. Equation (11) gives

$$\text{Var}\,X = \frac{N\,Q}{P}(\sigma_x^2 + a^2) , \tag{30}$$

$$\text{Var}\,Y = \frac{N\,Q}{P}(\sigma_y^2 + b^2) , \tag{31}$$

$$\text{Var}\frac{X}{Y} = \text{Var}\frac{\Sigma\, a_i\, \delta_i}{\Sigma\, b_i\, \delta_i} . \tag{32}$$

The approximation written as Equation (26) now leads to

$$\text{Rel-Var}\,\frac{X}{Y} = C_X^2 + C_Y^2 - 2\,C_{XY}\,. \tag{33}$$

The rel-variances C_X^2 and C_Y^2 have been conquered, but we have yet to evaluate C_{XY}. First, by definition

$$\text{Cov}\,X,\,Y = E\left[\sum \frac{a_i\,\delta_i}{P} - E\sum \frac{a_i\,\delta_i}{P}\right]\left[\sum \frac{b_i\,\delta_i}{P} - E\sum \frac{b_i\,\delta_i}{P}\right]$$

$$= E\left[\sum a_i\left(\frac{\delta_i}{P} - 1\right)\right]\left[\sum b_i\left(\frac{\delta_i}{P} - 1\right)\right]$$

[all the sums run over $i = 1$ to $i = N$]

$$= E\sum a_i\,b_i\left(\frac{\delta_i}{P} - 1\right)^2 + E\sum_{j \neq i} a_i\,b_j\left(\frac{\delta_i}{P} - 1\right)\left(\frac{\delta_j}{P} - 1\right)$$

$$= \sum a_i\,b_i\,E\left(\frac{\delta_i^2}{P^2} - \frac{2\,\delta_i}{P + 1}\right) + 0$$

$$= \sum a_i\,b_i\left(\frac{P}{P^2} - \frac{2\,P}{P + 1}\right) = \frac{Q}{P}\sum a_i\,b_i$$

$$= \frac{N\,Q}{P}\,a\,b + \frac{N\,Q}{P}\,\varrho\,a\,b\,C_x\,C_y\,, \tag{34}$$

wherein ϱ, C_x, and C_y have already been defined. We return now to Equation (33) for Plan II, whence

$$\text{Rel-Var}\,\frac{X}{Y} \doteq C_X^2 + C_Y^2 - 2\,C_{XY} \quad \text{[Equation (33)]}$$

$$= \frac{\text{Var}\,X}{(E\,X)^2} + \frac{\text{Var}\,Y}{(E\,Y)^2} - 2\,\frac{\text{Cov}\,X,\,Y}{EX\,EY}$$

$$= \frac{N\,Q}{P(N\,a)^2}\,(\sigma_x^2 + a^2) + \frac{N\,Q}{P(N\,b)^2}\,(\sigma_y^2 + b^2)$$

$$- 2\,\frac{N\,Q}{P\,N^2\,a\,b}\,(a\,b + \varrho\,a\,b\,C_x\,C_y)$$

$$= \frac{Q}{N\,P}\,(C_x^2 + C_y^2 - 2\,\varrho\,C_x\,C_y)\,, \tag{35}$$

which after replacement of NP by $E\,\hat{n} = n$ appears to be precisely what we wrote in Equation 26 for Plan I. Thus, although Plan II may show a severe loss of precision for the estimates X and Y of the total x- and y-populations in the frame, it is equivalent to Plan I for the ratio X/Y.

We note in passing that, by algebraic rearrangement, Equations (26) and (35) may be written as

$$\text{Rel-Var}\,\frac{X}{Y} \doteq \left(\frac{1}{n} - \frac{1}{N}\right)\frac{1}{N}\sum\left[\frac{a_i - \varphi\, b_i}{a}\right]^2. \tag{36}$$

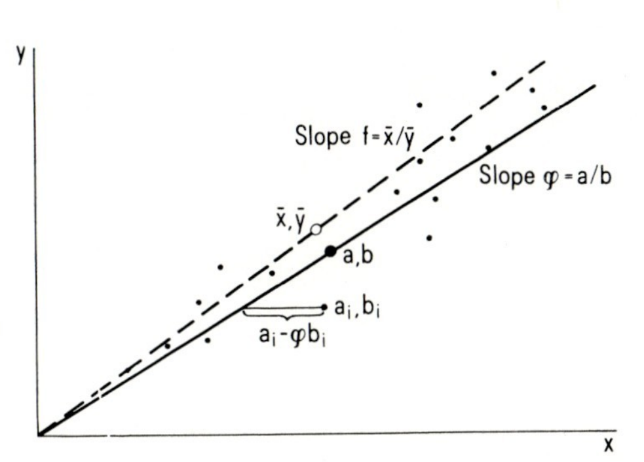

The complete coverage of the frame would give the centroid a, b. A line through the centroid and the origin would have slope $\varphi = a/b$. The sample of points has the centroid $\bar{x}/\bar{y}$. The line that connects it with the origin has slope $f = \bar{x}/y$.

The factor

$$\frac{1}{N}\sum\left[\frac{a_i - \varphi\, b_i}{a}\right]^2$$

is the average square of the vertical deviations of the N points a_i, b_i from the line $x = \varphi\, y$, measured in units of a, where $\varphi = A/B = a/b$. The sample-analog

$$\text{Rel-}\widehat{\text{Var}}f = \text{Rel-}\widehat{\text{Var}}\,\frac{X}{Y} \doteq \left(\frac{1}{n} - \frac{1}{N}\right)\frac{1}{n\,\bar{x}^2}\sum_1^n (x_i - f\,y_i)^2 \tag{37}$$

may be used as an estimator of the rel-variance of X/Y, though we usually replace $n\,\bar{x}^2$ by $(n-1)\,\bar{x}^2$.

4. Effect of Blanks in the Frame

Illustration from practice

It often happens in practice that one wishes to estimate the aggregate value of some characteristic of a subclass when the total number of units in the subclass is unknown. For example, in a study of consumer reserarch there was need for an estimate of the number of women aged 30 or over that live in a certain district, with no child under 12 years old at home: also the total disposable income of these women.

Suppose for simplicity that the frame is a list of all the occupied dwelling units in the district. Our sample will be a simple random sample of n dwelling units, drawn without replacement by reading out n random numbers between 1 and N, where N is the total number of dwelling units in the district. We depart for convenience from the notation at the front and use the subscript 1 for the specified subclass. Information is obtained on the n dwelling units in the sample, and it is noted that $\hat{n}_1$ is the count of dwelling units in this sample that contain women that belong to the specified subclass—that is, females 30 or over with no child under 12. Let $\bar{x}_1$ be the average income per dwelling unit in these $\hat{n}_1$ dwelling units. Some incomes in the specified subclass may be 0. $\hat{n}_1$ and $\bar{x}_1$ are both random variables: so is their product $\hat{n}_1 \bar{x}_1$, the total income of the women in the sample that belong to the specified subclass.

The reader will recognize the above sampling procedure as Plan I. We encounter in practice two main problems:

Problem 1. What is the variance of a ratio such as $\bar{x}_1$?

Problem 2. What is the variance of an estimator of a total, such as the total number of women in the subclass, or their total income?

We note first that the conditional expected value of $\bar{x}_1$ over all the samples that have n_1 dwelling units that contain women that belong to the specified subclass has the convenient property of being the average income of all the women in the frame that belong to the subclass. It is for this reason that the conditional rel-variance of $\bar{x}_1$ is useful for assessing the precision of a sample at hand.

What is the rel-variance of $\bar{x}_1$? Let C_1^2 be the rel-variance of incomes between the dwelling units in the frame that belong to the specified subclass. It is a fact that for the plan of sampling described here, the conditional rel-variance of $\bar{x}_1$, for samples of size $\hat{n}_1$ of the specified subclass, will be very nearly

$$\text{Rel-Var}\,\bar{x}_1 = \left(1 - \frac{n}{N}\right) \frac{C_1^2}{\hat{n}_1} \tag{38}$$

as if the dwelling units of this subclass in the frame had been set off beforehand in a separate stratum (Stratum 1), and a sample of size $\hat{n}_1$ drawn therefrom.

Incidentally, in the design-stage, for calculation of the size of sample to meet a prescribed $\text{Var}\,\bar{x}_1$, one may speculate on a value of P for the proportion of women in the frame that belong to the specified subclass, and then for the sampling procedure described above calculate the required size n of the sample by use of the formula

$$\text{Av}\,\text{Var}\,\bar{x}_1 = C_1^2\, E\, \frac{1}{\hat{n}_1} \doteq \frac{C_1^2}{n\,P} \left(1 + \frac{Q}{n\,P}\right). \tag{39}$$

The term Q/nP in the parenthesis arises from the fact that $\hat{n}_1$ is unpredictable, being a random variable.

We now turn our attention to Problem 2, estimation of the total number of women in the subclass. An estimator of X_1 will be

$$X_1 = \frac{N}{n} \hat{n}_1 \bar{x}_1 \tag{40}$$

using N/n as an expansion factor. If we place $x_i = 1$ for dwelling unit i in the sample if it contains a woman in the specified subclass, and place $x_i = 0$ otherwise, then X_1 will be an estimate of the number of women in the frame that belong to the specified subclass. We note immediately that, as N and n are known (not random), the conditional rel-variance of X_1 for samples of size $\hat{n}_1$ is exactly equal to the rel-variance of $\bar{x}_1$.

Unfortunately, though, this estimator X_1 of the total number or total income of all the women in the frame that belong to the specified subclass will not have all the convenient properties of $\bar{x}_1$. Thus, the conditional expectation of $X_1 = \hat{n}_1 E\, \bar{x}_1$ for samples that contain $\hat{n}_1$ members of the specified subclass is not equal to the aggregate income of all the women in the frame that belong to the subclass. One must conclude that the conditional rel-variance of X_1 for a sample at hand, although equal to the conditional rel-variance of $\bar{x}_1$, requires careful interpretation.

Instead of attempting to interpret the conditional variance of X_1, we may turn our attention to the average variance of X_1 in all possible samples of size n. We need more symbols. Subscript 1 will refer to the specified subclass; subscript 2 to the remainder. The word income will hereafter mean income from women of the specified subclass.

a_1	the average income per dwelling unit in the frame for the women that belong to the specified subclass. (Some incomes may of course be 0 in this subclass.)
$a_2 = 0$	for the remainder, because every sampling unit not in the specified subclass is a blank.
P	the proportion of the dwelling units in the frame that contain women of the specified subclass. $Q = 1 - P$.
σ_1^2	the variance in incomes between the dwelling units in the frame that belong to the specified subclass.
$a = P\, a_1$	the overall average income per dwelling unit in the frame, for the women of the specified subclass, including blanks (dwelling units with no women of the specified subclass).

We note that the overall variance between the incomes in all N dwelling units of the frame will be

$$\begin{aligned} \sigma^2 &= P\,\sigma_1^2 + Q\,\sigma_2^2 + P\,Q(a_2 - a_1)^2 \quad [\sigma_2 = 0] \\ &= P(\sigma_1^2 + Q\,a_1^2) = P\,a_1^2(C_1^2 + Q)\,. \end{aligned} \tag{41}$$

To find the average $\operatorname{Var} X_1$ over all possible samples, we may then use Equation (5), which gives

$$\operatorname{Var} X_1 = \left(1 - \frac{n}{N}\right) N^2 \sigma^2 / n$$

$$= \left(1 - \frac{n}{N}\right) \frac{N^2 P a_1^2 (C_1^2 + Q)}{n} \tag{42}$$

or in terms of rel-variance,

$$\text{Rel-Var } \bar{x}_1 = \left(1 - \frac{n}{N}\right) \frac{C_1^2 + Q}{n P} \tag{43}$$

in which we recognize $n P$ as $E \hat{n}_1$. The average variance of X_1 is thus afflicted by the proportion Q of blanks, whereas the average variance of $\bar{x}_1$ in Equation (38) is not.

As the proportion of blanks Q increases toward unity, Plan I becomes more and more the equivalent of Plan II with the same probability P of selection.

The problem with the variance of X_1 in Plan II arose from the assumption that N_1, the number of dwelling units in the frame with women that meet the specification of the subclass, is unknown. If N_1 were known, as sometimes it is, one could form $\bar{x}_1$ from the sample and then use the estimator

$$N_1 = N_1 \bar{x}_1 \tag{44}$$

which would have all the desirable properties of $\bar{x}_1$.

This observation suggests use of a preliminary sample by which to estimate the proportion of the total frame that belongs to the specified class. Briefly, the procedure is this: (1) to select from the frame by random numbers a preliminary sample of sufficient size N'; (2) to classify into strata by an inexpensive investigation, the units of the preliminary sample; (3) to investigate samples of sizes $\hat{n}_1$ and $\hat{n}_2$ from the two strata, to acquire the desired information. The preliminary sample furnishes estimates $\hat{P}_1$ and P_2 of the proportions of the two strata, and the final sample gives $\bar{x}_1$ and $\bar{x}_2$. The estimator is

$$\bar{x} = \hat{P}_1 \bar{x}_1 + \hat{P}_2 \bar{x}_2 . \tag{45}$$

The final sample may be selected proportionately from the strata of the preliminary sample, or (where advantageous) by Neyman allocation.

If the sorting into strata is successful, then the sample from Stratum 2 can be relatively small. It is in practice risky to reduce it to 0 for the simple reason that in most experience a few false positives in Stratum 2 are very effective in increasing the variance of $\bar{x}$.

Approximate variances for the two allocations are

$$\operatorname{Var} \bar{x} = \frac{\sigma_b^2}{N'} + \frac{\sigma_w^2}{n} \quad \text{[Proportionate allocation]} \tag{46}$$

and

$$\operatorname{Var} \bar{x} = \frac{\sigma_b^2}{N'} + \frac{(\bar{\sigma}_w^2)}{n}, \quad \text{[Neyman allocation]} \tag{47}$$

where σ_w^2 is the usual weighted average variance between sampling units within strata, and $\bar{\sigma}_w$ is the weighted average standard deviation between sampling units within strata. σ_b^2 is the variance between the means of ths strata.

There is an optimum size for the preliminary sample given by

$$\frac{n}{N'} = \frac{\sigma_w}{\sigma_b} \sqrt{\frac{c_1}{c_2}}, \quad \text{[Proportionate allocation]} \tag{48}$$

$$\frac{n}{N'} = \frac{\bar{\sigma}_w}{\sigma_b} \sqrt{\frac{c_1}{c_2}}, \quad \text{[Neyman allocation]} \tag{49}$$

where c_1 is the average cost to classify a sampling unit into a stratum, and c_2 is the average cost to investigate a unit in the final sample.

The theory is well known and need not be elaborated here. Such problems are complicated by the fact that estimation for several subclasses may be required in the same study.

Examples of blanks in the frame will be found in almost any book on sampling, one of the best being Chapter 9 in the 3rd edition of Frank Yates, *Sampling Methods for Censuses and Surveys* (Griffin, 1971). An example of calculations for a choice between Plans I and III appears in the author's book *Sample Design in Business Research* (Wiley, 1960), page 129.

We end on a further note of possible interest. If all the a_i in the specified subclass take the value 1, then $\sigma_1^2 = 0$ in Equation (41). Suppose now that the proportion Q of blanks approaches 1 and that n increases in a manner that holds $n\,P = m$. This circumstance corresponds to a count of flaws in test-panels of fixed size (fixed n, as of paint, or of a textile) in which the number of flaws in a test-panel may for practical purposes be infinite, but with an expected value of m. Equation (43) then leads to the Poisson

$$\text{Rel-Var}\, \hat{m} \to \left(1 - \frac{n}{N}\right) \frac{1}{m}, \tag{50}$$

n/N being the proportion of all panels that are observed.

5. Appendix: Illustration of Plan II with a Frame of Two Units

Table 3
The frame.

Serial numbers of sampling unit	Populations	
	x	y
1	$a_1 = 1$	$b_1 = 3$
2	$a_2 = 2$	$b_2 = 5$

Table 4
Statistical properties of the frame.

Total population	$A = 3$	$B = 8$
Average per sampling unit	$a = 1.5$	$b = 4$
Standard deviation	$\sigma_x = 1/2$	$\sigma_y = 1$
Coefficient of variation	$C_x = 1/3$	$C_y = 1/4$

$\varphi = A/B = a/b = 3/8 = .375,$

$\text{Cov}\, x, y = \frac{1}{2}\,(.5 \cdot 1 + .5 \cdot 1) = 1/2,$

$\varrho = \text{Cov}\, x, y/\sigma_x\,\sigma_y = \frac{1}{2}/\frac{1}{2} \cdot 1 = 1$ (always true with 2 points),

$C_{xy} = \text{Cov}\, x, y/a\,b = \frac{1}{2}/1.5 \cdot 4 = 1/12.$

We now list the 4 possible outcomes of the sampling procedure for $P = Q = 1/2$. Their expected proportions are equal. We observe that:

1. $E\,\hat{n} = \frac{1}{4}\,(0 + 1 + 1 + 2) = 1.$ $NP = 2 \cdot 1/2 = 1$, in agreement.

2. Av $X = 3 = A$, which illustrates the unbiased character of the sampling procedure. Likewise, Av $Y = 8 = B$.

3. $\text{Var}\,X = \frac{1}{4}\{(0-3)^2 + (4-3)^2 + (2-3)^2 + (6-3)^2\} = 20/4 = 5.$

In comparison, the formula for Var X gives

$$\text{Var}\,X = (N\,Q/P)\,(\sigma_x^2 + a^2)$$

$$= 2(\tfrac{1}{4} + 1.5^2) = 20/4 = 5\,.$$

Obviously, most of this variance comes from the term $a^2 = 1.5^2$.

Table 5
Table of all possible samples selected from Plan II from the frame shown above, with $P = Q = 1/2$.

Sampling units in sample	x-population of sample x	y-population of sample y	$X = 2x$	$Y = 2y$	X/Y	x
Both out	0	0	0	0	–	–
No. 1 out, No. 2 in	2	5	4	10	4/10	2
No. 1 in, No. 2 out	1	3	2	6	2/6	1
Both in	3	8	6	16	6/16	1.5
Average	1.5	4	3	8	133/360	1.5

4. Suppose that we know N, and that we use the estimator

$$X' = N\bar{x} = 2\bar{x}$$

for the total x-population. The three useable values of X' would then be 4, 2, 3, whose average value agrees with $A = 3$. We note that

$$\operatorname{Var} X' = \frac{1}{3}[(4-3)^2 + (2-3)^2 + (3-3)^2]$$

$$= \frac{2}{3}$$

which is much less than $\operatorname{Var} X = 5$, just encountered. $\operatorname{Var} X'$ has all the desirable properties of $\bar{x}$.

5. Every sample, if it contains a sampling unit, gives an estimate of $\varphi = A/B = 3/8 = .375$. The 3 possible estimates are in the table. Their average is $133/360 = .3694$. The sampling procedure for estimation of φ is therefore slightly biased, as statistical theory would lead us to expect. The bias is incidentally $.3694 - .3750 = .0056$, being only 15 parts in 1000, or only 4.7% of the standard error of X/Y.

6. Equation (34) gives the approximation

$$C^2_{X/Y} \doteq \frac{1}{E\,\hat{n}}[C^2_x + C^2_y - 2\,C_{xy}]$$

$$= \frac{1}{1}\left[\left(\frac{1}{3}\right)^2 + \left(\frac{1}{4}\right)^2 - 2/12\right]$$

$$= \frac{1}{9} + \frac{1}{16} - \frac{2}{12} = \frac{1}{144} = .006\,944\,.$$

7. The table of all possible samples gives

$$\sigma^2_{X/Y} = \frac{1}{3}[(4/10 - 133/360)^2 + (2/6 - 133/360)^2 + (6/16 - 133/360)^2]$$

$$= .002\,862/3$$

$$= .000\,954\,3$$

$$C^2_{X/Y} = .000\,954\,3/(133/360)^2$$

$$= .006\,991\,7$$

in closer agreement with .006944 than we might expect for samples of $n = 1$.

Acknowledgement

I am deeply indebted to my friend and colleague Dr. Morris H. Hansen for calling my attention years ago to Plan II, and for his continued interest and help in the theory and comparison between Plans I, II, and III in practice. I have already expressed my indebtedness to William N. Hurwitz. It is a pleasure to mention also my good fortune to work with Professor William H. Kruskal on the problem of blanks, during the preparation of my article Survey Sampling for the New Encyclopedia of the Social Sciences.

Author's address:
W. Edwards Deming, 4924 Butterworth Place, Washington 20016.

I. Higuti

I. Higuti

Some Remarks on the Concepts of Size-Distribution in the Study of Small Particles

Summary

The basic concepts of size-distribution in the study of small particles are discussed in a rather abstract version. The conventional number- and weight-distributions are looked at from a more general point of view, making clear the relation between them and elucidating other sorts of distribution concept.

A general view of method of measurements is also described, classifying various methods into groups according to the extent of separation in the operation of measurement, so that they correspond closely to the basic concepts discussed.

1. **Introduction**

The concept of distribution in statistics is usually based on the measurements of individuals in the ensembles to be studied. For example, in the study of physical characteristics of people in a society, we measure the height of each individual at first, enumerate the number of observations which belongs to respective predesignated classes and then we build up a distribution of the height of people in the society. In the study of particular materials, however, we often come across circumstances where it is impossible, or tremendously laborious, to measure individuals, while to measure an arbitrarily chosen subset of the ensemble as a whole is comparatively easy. So in small particle statistics a special concept of distribution has been developed, namely the concept of weight-distribution. Therefore we have two kinds of distribution—i.e. the number-distribution (which is the ordinary one in statistics) and the weight-distribution.

Are these the only two distributions?

The answer is no; people are occasionally measuring other kinds and quite unconsciously they think that what they are handling is one of these two.

The aim of this short article is to make clear the basic concepts of various size-distributions in small particle statistics and then the methods of their measurement, as well as the relations between them, and to suggest the possibility of building up new concepts of distribution and their measurement.

2. Distributions of Various Sorts

Consider a set S of N particles, $A_1, A_2, \ldots, A_N$. Generally, the statistical property of S based on the property of individuals is represented by an average

$$\mathrm{Av}\{F(X, Y, \ldots, Z)\} = \frac{1}{N}\sum_{i=1}^{N} F\big(X(A_i), \ldots, Z(A_i)\big)$$

of a function $F(X, Y, \ldots, Z)$ of the property $X, \ldots, Z$ with respect to $A_1, A_2, \ldots, A_N$ or by a quantity which is derived from some averages of the same kind. Of course, the extensive properties of S, such as total number, total weight, etc., are not average. But the particle size distributions are intensive properties and can be thought of as averages. In the sequel we shall restrict ourselves to particle size which assumes a definite value for each particle and we shall designate them as X (e.g. the maximum diameter, the volume, etc.). The mean size of the particles is then the average of X itself: $\mathrm{Av}\{X\}$.

The number distribution (cumulative distribution function) of X is then the average of a function of X with a parameter x, which is represented by means of unit step function $D(x)$ as

$$F(X; x) = D\big(x - X(A_i)\big).$$

In small particle statistics the distribution of the following type also appears

$$\phi(x; X \mid Y) = \mathrm{Av}\{D(x - X)\, Y\}/\mathrm{Av}\{Y\},$$

where X denotes any quantity which represents the dimension of particles and Y is any quantity which is additive with respect of subsets of S. (A single particle is also looked on as a subset of S. By additivity we mean for subset S_i which consists of particles $A_{i1}, A_{i2}, \ldots, A_{ik}$, the relation

$$Y(S_i) = \sum_{j=1}^{k} Y(A_{ij})$$

holds.)

The volume is additive, but the diameter is not. Y should have, of course, a certain physical meaning. In case Y is the weight W, $\phi(x; X \mid W)$ is so-called weight-distribution of X. There can be various weight-distributions according to the choice of X; weight-distribution of maximum diameter, weight-distribution of surface area, etc., as is well known.

As a special case, if $Y(A_i)$ is independent of A_i, i.e. for any A_i it is the same value, and therefore $Y(S_i)$ is only proportional to the number of particles, the function $\phi(x; X \mid Y)$ is nothing else than the number distribution of X. In this sense we can say that $\phi(x; X \mid Y)$ is more general than the ordinary distribution concept.

Our proposition is now that the quantity Y in $\phi(x; X \mid Y)$ need be neither weight nor number.

The area M of the shadow of the dispersed particles in a solution may be assumed to be additive, though strictly not so if orientations of particles are taken into account. Therefore, it is quite right to speak of a size-distribution in which M is adopted as Y. What is directly measured in the mearurement of particle-size distribution of small particles in a suspension by the light extinction method is $\phi(x; X \mid M)$ where X is the Stokes' diameter of the particles.

Now we have treated only the cumulative version of distribution. Practically useful density functions, of course, don't exist since the particles are discrete. But if we think of distribution functions as smooth as differentiable, and if we differentiate them, we have densities. Using formal calculation, changing the order of differentiating and averaging we have expressions, which contain Dirac's delta function, which in turn prove convenient for further calculations.

Since there are many distributions the calculating method of conversion from one to another is often required. In many cases it requires more assumptions for particles, namely we must replace particles by ideal ones. To discuss them from a general point of view is of some interest but we cannot go into detail now.

3. Basic Ideas of Measuring Size-Distributions of Particles

From a statistical point of view, the methods of measuring particle-size distributions of particles of an ensemble may be classified into the following four types:

(i) method of individual measurements,
(ii) method of complete separation,
(iii) method of incomplete separation,
(iv) method of complete non-separation.

The method of individual measurement is simply based on the observations on individuals, and is very suitable for the evaluation of number-distributions. The measurement of size distribution from the microscopic photograph belongs to this group.

The typical practical technique which corresponds to the method of complete separation is sieving. We divide the set S into subsets $S_1, S_2, \ldots, S_n$ by size, so that each S_i consisting of only those particles A_i whose size characteristics $X(A_i)$ lie on a definite interval (x_{i-1}, x_i). By this method the histogram of the distribution $\phi(x; X \mid Y)$ is measured.

The method of incomplete separation, an indirect method often used for fine particles, is based on the measurement of $Y(S'_j)$, where S'_j is a set which contains the subset S_j of S consisting of all A_i with $X(A_i) < x_j$. In case S'_j coincides with its corresponding S_j the treatment can be reduced to the case of complete separation owing to the additivity of Y. Otherwise we may need to consider the following points.

In many cases a set which contains $S - S_j$, say S''_j, is formed instead of S'_j.

In practice $Y(S)$, $Y(S''_1)$, $Y(S''_2), \ldots$ or the ratios of $Y(S''_j)$ to $Y(S)$ is observed. It suffices to obtain $Y(S_j)$ from these observations, because then the method can be reduced to the case of complete separation. Let

$$S''_j = (S - S_j) + R_j ,$$

then it is easily seen from the additivity of Y that

$$Y(S) - Y(S''_j) = Y(S_j - R_j)$$

holds. Therefore there would be no problem, if $Y(S_i - R_j)/Y(S_j)$ were elucidated as a function of x. This approach, however, is fruitless. If, however, for each individual particle, whether it belongs to R_j or not, and is quite independent of other particles, then the expectation of $Y(S_j - R_j)$ in this probability field becomes

$$\sum_{i=1}^{N} \big(1 - \Pr(A_i \,\varepsilon\, R_j)\big) D\big(x_j - X(A_i)\big) Y(A_i) .$$

Therefore if $\Pr(A_i \,\varepsilon\, R_j)$ is known as a function $p(X)$ which depends only on the value of $X(A_i)$, it becomes

$$\frac{\text{expectation of } Y(S_j - R_j)}{Y(S)} = \frac{\mathrm{Av}\{\big(1 - p(X)\big) D(x_j - X)\, Y\}}{\mathrm{Av}\{Y\}} .$$

Replacing the left hand side of this formula by observations

$$1 - \frac{Y(S''_j)}{Y(S)} ,$$

we easily get a method of estimating the distribution, namely $\phi(x\colon X \mid Y)$, from observations of $Y(S''_j)/Y(S)$ for many S''_j we construct a continouus function

$$Y\big(S''(x)\big)/Y(S)$$

of x. Then for this function the above formula becomes

$$1 - \frac{Y(S''(x))}{Y(S)} = \int_0^x (1 - p(x))\, d\phi(x; X \mid Y) .$$

This is a general formula for the measurement of $\phi(x; X \mid Y)$ by the method of incomplete separation containing the method of separate precipitation (both gravitational and centrifugal). If $p(x)$ is known theoretically from any physical consideration we can obtain the distribution by solving numerically this integral equation.

If we have some methods of exciting particles whose responses are sensitive to the particle size, we may use it, in some cases, to measure the particle size distribution without any actual sorting of particles. We call these methods of complete non-separation. In these methods changing the magnitude of excitation we observe the variation of response intensity. To be able to get size distribution the response characteristic should be linear and additive.

Sonic method, the method of retardation spectra, method of liquid double refraction are examples of some of the methods of the complete non-separation technique.

In case of double refraction the ratio of two different responses is observed. Let $R_1(S, \alpha)$, $R_2(S, \alpha)$ be two responses, then

$$\begin{aligned} R_l(S, \alpha) &= \mathrm{Av}\{R_l(A, \alpha)\} \\ &= \int \phi_l(x, \alpha)\, dF_x(x) \quad (l = 1, 2) , \end{aligned}$$

where $\phi(x, \alpha)$ is the response of the sub-ensemble which consists of particles with $X = x$. Theoretically $F(x; X)$ can be obtained from observations of $R_1(S, \alpha)/R_2(S, \alpha)$ for various α, but practically it is very difficult to get steady solution numerically. If $\phi_l(x, \alpha)$ attains 0 for any x more than $x_0(\alpha)$ which is determined by α it becomes the case which can be reduced to the case of complete separation.

It is quite conceivable that these methods are superior to the case of non-separation, if technical difficulties are comparable.

According to this idea a method is developed by H. Noda and H. Funakoshi which uses rotation electromagnetic field for polarised particles of various lengths.

4. Conclusions and Acknowledgement

In this short article only the surface of concepts and methods of measuring various size-distributions has been scratched. It might be thought that such a rather abstract discussion contributes nothing to the practical problems. But the author has experienced a fairly long sterile discussion among the practicians on the effect of the photo-absorption coefficient on the observed distribution by means of the light extinction method only because of their lack of basic understanding of the distribution being measured.

The author is obliged to Mr. J. Nicholas who helped in the final drafting of the manuscript.

References

[1] HERDAN, G. and SMITH, M. (1953): Small particle statistics. Elsevier Co.
[2] DALLAVAL, J. M. (1948): Micromeritics. Pitman Pub.
[3] HIGUTI, I. (1964): Some studies in particle statistics (Japanese). Proceedings of the Institute of Statistical Mathematics. Vol. *12*, No. 1.
[4] HIGUTI, I. (1956): A remark on Tromp-Curve Analysis (Japanese). Proceedings of the Institute of Statistical Mathematics. Vol. *3*, No. 2.

Author's address:
Dr. rel. nat. Isao Higuti, The Institute of Statistical Mathematics, 4–6–7 Minami-Azabu Minatoku, Tokyo, Japan.

K. Stange[1] †

K. Stange

Der Einfluss der Genauigkeit des Messgerätes beim «Auslesen» nicht-masshaltiger Teile aus einer Fertigung

1. Aufgabenstellung

Bei einer Fertigung seien die Merkmalwerte x (nahezu) normal verteilt mit der Dichte $\psi(x \mid \mu; \sigma_x^2) = \psi(x)$, wobei μ und σ_x^2 Mittelwert und Varianz von x bedeuten,

$$M(x) = \mu \quad \text{und} \quad V(x) = \sigma_x^2 . \tag{1.1}$$

Für die Merkmalwerte x sei *eine* Toleranzgrenze T vorgeschrieben. Ohne die Allgemeinheit der Untersuchung einzuschränken, wird T als *untere* Toleranzgrenze gewählt, das heisst alle x-Werte mit $x < T$ gelten als «schlecht» und sollen ausgelesen werden. Alle x-Werte mit $x \geqslant T$ gelten als «gut». Der Schlechtanteil der Fertigung ist

$$p_x = \int_{-\infty}^{T} \psi(x)\, dx = \Psi(T) . \tag{1.2}$$

Es sei $\Psi(T) \neq 0$. Beim Auslesevorgang werden die (wahren) Merkmalwerte x mit einem Messgerät beurteilt. Da kein Messgerät fehlerfrei arbeitet, werden gelegentlich schlechte x-Werte ($x < T$) als «gut» und gute x-Werte ($x \geqslant T$) als «schlecht» eingestuft, wie man sich am Sonderfall $x = T$ in Fig. 1 sehr anschaulich klar macht.

Die «Fehler» ε des benutzten Messgeräts seien normal verteilt mit

$$M(\varepsilon) = 0 \quad \text{und} \quad V(\varepsilon) = \sigma_\varepsilon^2 . \tag{1.3}$$

Wegen $M(\varepsilon) = 0$ hat das Gerät keine systematische Abweichung. Die durch $1/V(\varepsilon)$ gekennzeichnete Messgenauigkeit ist (mindestens im x-Bereich der Fer-

[1]) Gestorben am 23. Juni 1974. Die Korrektur des Fahnenabzuges dieses Beitrages übernahm für seinen Lehrer in verdankenswerter Weise Herr Dr. H. Hilden, Basel.

tigung) unabhängig von x. Dem wahren Wert x wird der Messwert y zugeordnet. Aus

$$y = x + \varepsilon \tag{1.4}$$

folgt

$$M(y) = \mu \quad \text{und} \quad V(y) = \sigma_y^2 = \sigma_x^2 + \sigma_\varepsilon^2 . \tag{1.5}$$

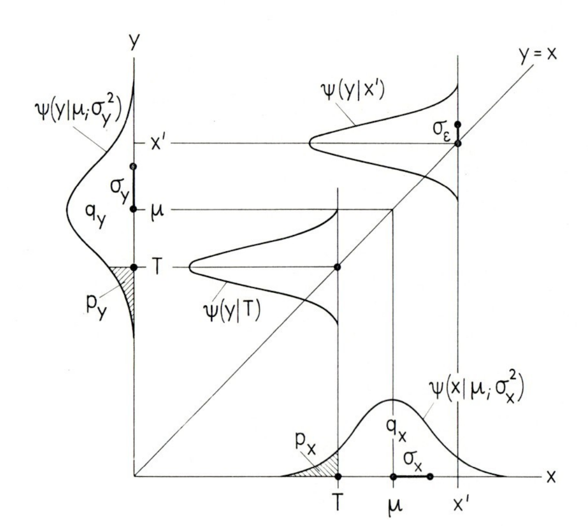

Fig. 1
Die Randverteilung der Fertigungswerte x mit $(\mu;\, \sigma_x^2)$, die Verteilung der Messfehler ε mit $(0;\, \sigma_\varepsilon^2)$ und die Randverteilung der Messwerte y mit $(\mu;\, \sigma_x^2 + \sigma_\varepsilon^2)$.

Während der Mittelwert $M(y)$ der Messwerte gleich dem Mittelwert μ der Fertigung wird, ist die Varianz $V(y)$ der Messwerte grösser als die Varianz σ_x^2 der Fertigung. Dabei wird vorausgesetzt, dass die Mess«fehler» ε unabhängig von den Fertigungs«fehlern» $(x - \mu)$ sind. Der Einfluss des Mess- bzw. Ausleseverfahrens darf um so weniger vernachlässigt werden, je grösser die Messvarianz σ_ε^2 bei gegebener Fertigungsvarianz σ_x^2 ist. Der Einfluss von σ_ε^2 auf die «Güte des Auslesens» wird im folgenden untersucht.

2. Die zweidimensionale $(x; y)$-Verteilung

In Fig. 1 ist

$$\psi(x) = \frac{1}{\sqrt{2\pi}\,\sigma_x} \exp\left[-\frac{1}{2}\left(\frac{x-\mu}{\sigma_x}\right)^2\right] \tag{2.1}$$

die Randdichte der «gefertigten» x-Werte über der waagerechten x-Achse. Bei gegebenem x wird die bedingte Dichte der Messwerte y nach dem Vorausgehenden

$$\psi(y \mid x) = \frac{1}{\sqrt{2\pi}\,\sigma_\varepsilon} \exp\left[-\frac{1}{2}\left(\frac{y-x}{\sigma_\varepsilon}\right)^2\right]. \tag{2.2}$$

Die Randdichte $\psi(x)$ und die bedingte Dichte $\psi(y \mid x)$ bestimmen die Gestalt der zweidimensionalen $(x; y)$-Verteilung mit der Dichte $\psi(x)\,\psi(y \mid x)$ vollständig. Da sowohl $\psi(x)$ als auch $\psi(y \mid x)$ Dichten von Normalverteilungen sind, genügt das Wertepaar $(x; y)$ einer zweidimensionalen Normalverteilung. Ihre (normale) Randdichte über der senkrechten y-Achse ist nach (1.5) $\psi(y) = \psi(y \mid \mu; \sigma_x^2 + \sigma_\varepsilon^2)$.

Es sei $\varrho(x; y) = \varrho$ die Korrelationszahl zwischen x und y. Aus der bekannten Beziehung für die bedingten Mittelwerte von y,

$$M(y \mid x) = \mu_y + \varrho(\sigma_y/\sigma_x)\,(x - \mu_x)\,, \tag{2.3}$$

folgt hier mit $\mu_x = \mu_y = \mu$ und $\mathrm{M}(y \mid x) = x$ leicht

$$x = \mu + \varrho(\sigma_y/\sigma_x)\,(x - \mu)\,. \tag{2.4}$$

Danach wird die Korrelationszahl ϱ

$$\varrho = \sigma_x/\sigma_y = \frac{1}{\sqrt{1 + (\sigma_\varepsilon/\sigma_x)^2}} = \frac{1}{\sqrt{1+\lambda^2}} \quad \text{mit} \quad \lambda = \sigma_\varepsilon/\sigma_x\,. \tag{2.5}$$

Für $\sigma_\varepsilon \to 0$ bzw. $\lambda \to 0$ gilt $\varrho \to 1$, wie man es erwartet.

Der Anteil, der von der Gutseite der Fertigung mit $x \geqslant T$ beim Auslesen auf die «Schlechtseite» mit $y < T$ übertragen wird, sei α (Fehlaustrag von Gut nach «Schlecht»; Fig. 2).

	Beschaffenheit		
	schlecht $x < T$ bzw. $u < \tau$	gut $x \geq T$ bzw. $u \geq \tau$	
Beurteilung: „schlecht" $y < T$ bzw. $v < \tau$	$\Phi(\tau) - \beta$	$\alpha(\tau; \lambda)$	$\Phi(\tau) + \alpha - \beta = \Phi(\tau/\sqrt{1+\lambda^2}) = p_y$
Beurteilung: „gut" $y \geq T$ bzw. $v \geq \tau$	$\beta(\tau; \lambda)$	$1 - \Phi(\tau) - \alpha$	$1 - \Phi(\tau) - \alpha + \beta = 1 - \Phi(\tau/\sqrt{1+\lambda^2}) = q_y$
	$p_x = \Phi(\tau)$	$q_x = 1 - \Phi(\tau)$	1

Fig. 2
Beim Auslesen des Schlechtanteils $p_x = \Phi(\tau)$ einer Fertigung mit Hilfe eines Messgerätes entstehen die (nicht erwünschten) «Fehlausträge» α und β.

Es gilt

$$\alpha = \int_{x=T}^{\infty} \int_{y=-\infty}^{T} \psi(x)\, \psi(y \mid x)\, dx\, dy\,.$$

Die Integration über y ist mit (2.2) ausführbar und gibt

$$y = \int_{-\infty}^{T} \psi(y \mid x)\, dy = \Phi[(T - x)/\sigma_\varepsilon]\,,$$

wobei $\Phi(\ )$ die Summenfunktion der standardisierten Normalverteilung NV (0; 1) bedeutet. Damit wird

$$\alpha = \int_{x=T}^{\infty} \Phi[(T - x)\sigma_\varepsilon]\, \psi(x)\, dx = \alpha(\mu;\, \sigma_x^2;\, T;\, \sigma_\varepsilon^2)\,, \tag{2.6}$$

abhängig von den vier Parametern μ, σ_x^2, T und σ_ε^2. Der Anteil, der von der Schlechtseite der Fertigung mit $x < T$ beim Auslesen auf die «Gutseite» mit $y \geqslant T$ übertragen wird, sei β. (Fehlaustrag von Schlecht nach «Gut».) Es gilt

$$\beta = \int_{x=-\infty}^{T} \int_{y=T}^{\infty} \psi(x)\, \psi(y \mid x)\, dx\, dy\,.$$

Entsprechend wie bei α findet man hier

$$\beta = \int_{x=-\infty}^{T} \{1 - \Phi[(T - x)/\sigma_\varepsilon]\}\, \psi(x)\, dx = \beta(\mu;\, \sigma_x^2;\, T;\, \sigma_\varepsilon^2)\,. \tag{2.7}$$

3. Einführung von dimensionslosen Veränderlichen

Damit man von der Fertigungslage μ und den besonderen Masseinheiten des Merkmals x unabhängig wird, führt man die dimensionslosen Veränderlichen

$$(x - \mu)/\sigma_x = u \quad \text{und} \quad (y - \mu)/\sigma_x = v \tag{3.1}$$

ein. Dann gilt für Mittelwerte und Varianzen

$$M(u) = 0;\ V(u) = 1;\ M(v) = 0;\ V(v) = 1 + \lambda^2\,. \tag{3.2}$$

Während u standardisiert normal verteilt ist, gilt das gleiche für v wegen $\lambda \neq 0$ nicht.

Der Toleranzgrenze $x = T$ wird die standardisierte Grenze

$$(T - \mu)/\sigma_x = \tau \tag{3.3}$$

zugeordnet. Ferner wird das Argument von Φ in (2.6) und (2.7) zu

$$(T - x)/\sigma_\varepsilon = [(T - \mu + \mu - x)/\sigma_x] \cdot (\sigma_x/\sigma_\varepsilon) = (\tau - u)/\lambda \, . \tag{3.4}$$

Damit findet man α aus (2.6) in der Gestalt

$$\alpha = \int\limits_{u=\tau}^{\infty} \Phi[(\tau - u)/\lambda] \; \varphi(u) \; du = \alpha(\tau; \lambda) \, , \tag{3.5}$$

also nur noch abhängig von den zwei «wesentlichen» Parametern τ und λ; $\varphi(u)$ ist die Dichte der NV (0; 1).

Entsprechend wird β aus (2.7)

$$\beta = \int\limits_{u=-\infty}^{\tau} \{1 - \Phi[(\tau - u)/\lambda]\} \; \varphi(u) \; du = \beta(\tau; \lambda) \, . \tag{3.6}$$

Aus (3.5) und (3.6) sind die Anteile (α; β) für die Fehlausträge des Ausleseverfahrens (das heisst die falsch eingeordneten Anteile der Fertigung) berechenbar, und zwar in Abhängigkeit vom Wertepaar (τ; λ).

In praxi ist das Verhältnis der Varianzen $(\sigma_\varepsilon/\sigma_x)^2 = \lambda^2 < 1$, meist sogar erheblich kleiner als 1. Mindestens wird man Messgeräte bevorzugen, welche dieser Bedingung genügen. Dieser Sonderfall wird im folgenden weiterbehandelt.

4. Der Sonderfall $\lambda = \sigma_\varepsilon/\sigma_x \lessapprox 1/2$

In Gleichung (3.5) setzt man

$$(\tau - u)/\lambda = z \quad \text{bzw.} \quad u = \tau - \lambda \, z \, . \tag{4.1}$$

Dann wird

$$\alpha = \lambda \int\limits_{z=-\infty}^{0} \Phi(z) \; \varphi(\tau - \lambda \, z) \; dz \, . \tag{4.2}$$

Da man beim Auslesen im wesentlichen nur (dimensionslose) Merkmalwerte u «in der Nähe» der Toleranzgrenze τ falsch beurteilt, so liegt es nahe, in der letzten Gleichung $\varphi(u) = \varphi(\tau - \lambda z)$ an der Stelle $u = \tau$ bzw. $z = 0$ in die Taylor-Reihe

$$\varphi(u) = \varphi(\tau) - \lambda \varphi'(\tau) z + \lambda^2 \frac{\varphi''(\tau)}{2!} z^2 - \lambda^3 \frac{\varphi'''(\tau)}{3!} z^3 \pm \cdots$$

zu entwickeln. Da sich alle Ableitungen $\varphi^{(k)}(\tau)$ der (standardisierten) Normalverteilung in der Gestalt

$$\varphi^k(\tau) = \varphi(\tau) P_k(\tau)$$

ausdrücken lassen, wobei $P_k(\tau)$ (einfache) bekannte Polynome von τ sind, so wird die Reihenentwicklung

$$\varphi(u) = \varphi(\tau - \lambda z) = \varphi(\tau) \sum_{k=0}^{\infty} [(-1)^k/k!]\, \lambda^k P_k(\tau) z^k . \tag{4.3}$$

Setzt man den Ausdruck (4.3) in (4.2) ein und vertauscht die Reihenfolge von Integration und Summation, so findet man

$$\frac{\alpha}{\lambda \varphi(\tau)} = \sum_k (1/k!)\, \lambda^k P_k(\tau) (-1)^k \int_{-\infty}^{0} z^k \Phi(z)\, dz . \tag{4.4}$$

Entsprechend wird aus (3.6)

$$\frac{\beta}{\lambda \varphi(\tau)} = \sum_k [(-1)^k/k!]\, \lambda^k P_k(\tau) \int_{0}^{\infty} z^k [1 - \Phi(z)]\, dz . \tag{4.5}$$

Wie an anderer Stelle[1]) gezeigt wurde, gilt für das Integral in (4.4) bzw. (4.5)

$$J(k) = (-1)^k \int_{-\infty}^{0} z^k \Phi(z)\, dz = \int_{0}^{\infty} z^k [1 - \Phi(z)]\, dz$$

$$= \frac{1}{2(k+1)} \frac{2^{(k+1)/2}}{\sqrt{\pi}} \Gamma\left(\frac{k+2}{2}\right) . \tag{4.6}$$

[1]) K. Stange. Pläne für messende Prüfung bei Berücksichtigung von «Vorkenntnissen» über die Verteilung der Mittelwerte. Metrika 1974.

Damit ist die Reihenentwicklung von α bzw. β vollständig bekannt. Setzt man

$$P_0(\tau) = 1;\ P_1(\tau) = -\tau;\ P_2(\tau) = \tau^2 - 1;\ P_3(\tau) = -\tau(\tau^2 - 3);\ \ldots \quad (4.7)$$

(wobei man die P_k bedarfsweise aus der Rekursionsformel
$P_{k+1} = (dP_k/d\tau) - \tau P_k$ berechnet) und

$$J(0) = 1/\sqrt{2\pi};\ J(1) = 1/4;\ J(2) = (2/3)/\sqrt{2\pi};\ J(3) = 3/8;\ \ldots \quad (4.8)$$

nach (4.6) in die Gleichung (4.4) ein, so wird ausführlich aufgeschrieben

$$\alpha(\tau;\lambda) = \lambda\,\varphi(\tau)\left[\frac{1}{\sqrt{2\pi}} - \frac{\tau}{4}\lambda + \frac{\tau^2 - 1}{3\sqrt{2\pi}}\lambda^2 - \frac{\tau(\tau^2 - 3)}{16}\lambda^3 + \cdots\right]. \quad (4.9)$$

Entsprechend findet man aus (4.5)

$$\beta(\tau;\lambda) = \lambda\,\varphi(\tau)\left[\frac{1}{\sqrt{2\pi}} + \frac{\tau}{4}\lambda + \frac{\tau^2 - 1}{3\sqrt{2\pi}}\lambda^2 + \frac{\tau(\tau^2 - 3)}{16}\lambda^3 + \cdots\right]. \quad (4.10)$$

Zur zahlenmässigen Berechnung bildet man zweckmässig die Summe und die Differenz der Fehlausträge. Es wird

$$\alpha + \beta = \sqrt{2/\pi}\,\lambda\,\varphi(\tau)\,[1 + (\lambda^2/3)\,(\tau^2 - 1)] \quad (4.11)$$

und

$$\alpha - \beta = (1/2)\,\lambda^2\,|\tau|\,\varphi(\tau)\,[1 + (\lambda^2/4)\,(\tau^2 - 3)]\,, \quad (4.12)$$

wobei das erste in der Klammer (!) gegen 1 vernachlässigte Glied in (4.11) bzw. (4.12)

$$(\lambda^4/15)\,(\tau^4 - 6\,\tau^2 + 3) \quad \text{bzw.} \quad (\lambda^4/24)\,(\tau^4 - 10\,\tau^2 + 15)$$

ist.

Die Differenz $(\alpha - \beta)$ lässt sich noch auf anderem Wege unabhängig von (4.12) berechnen. Da die transformierte Veränderliche v nach (3.2) mit $(0;\ 1 + \lambda^2)$ normal verteilt ist, so wird der als «schlecht» mit $y < T$ bzw. $v < \tau$ eingeordnete Anteil gleich $\Phi(\tau/\sqrt{1 + \lambda^2})$. Andererseits wird dieser Anteil nach Fig. 2 zu $\Phi(\tau) + \alpha - \beta$. Daraus folgt

$$\alpha - \beta = \Phi(\tau/\sqrt{1 + \lambda^2}) - \Phi(\tau)\,. \quad (4.13)$$

Entwickelt man

$$\Phi(\tau/\sqrt{\ }) = \Phi[\tau - (1/2)\,\lambda^2\,\tau + (3/8)\,\lambda^4\,\tau \mp \cdots] \equiv \Phi(\tau + \delta)$$

an der Stelle τ in eine nach Potenzen von δ fortschreitende Reihe, so findet man mit $d\Phi/d\tau = \varphi(\tau)$ und $d^2\Phi/d\tau^2 = -\,\tau\,\varphi(\tau)$ leicht

$$\Phi(\tau/\sqrt{\ }) = \Phi(\tau) - (1/2)\,\lambda^2\,\tau\,\varphi(\tau) + (1/8)\,\lambda^4\,\tau\,\varphi(\tau)\,(3 - \tau^2) + \cdots,$$

was mit (4.13) wiederum auf Gleichung (4.12) führt.

Mit Hilfe der Gleichungen (4.11) und (4.12) sind die Fehlausträge $(\alpha; \beta)$ des Ausleseverfahrens in Abhängigkeit von der (dimensionslosen) Toleranzgrenze τ und dem Verhältnis der Standardabweichungen $\lambda = \sigma_\varepsilon/\sigma_x \lessapprox 1/2$ berechenbar. Für $\tau \neq 0$ (was normalerweise der Fall ist) gilt nach (4.12) $\alpha > \beta$, was nach Fig. 1 auch anschaulich einleuchtet.

Für das Beispiel $\Phi(\tau) = 5\%$ (Schlechtanteil der Fertigung) mit $\tau = -\,1{,}645$ und $\varphi(\tau) = 0{,}1031$ findet man für $\lambda_1 = \sigma_\varepsilon/\sigma_x = 1/10$ bzw. $\lambda_2 = \sigma_\varepsilon/\sigma_x = 1/2$ die folgenden zwei Übersichten über die Fehlausträge (alle Zahlenwerte in %).

Beispiel 1 mit $\lambda = 1/10$	Beschaffenheit schlecht	 gut	
Beurteilung «schlecht»	4,63	α = 0,46	p_y = 5,09
Beurteilung «gut»	β = 0,37	94,54	q_y = 94,91
Alle Werte in %	p_x = 5,00	q_x = 95,00	100,00

Beispiel 2 mit $\lambda = 1/2$	Beschaffenheit schlecht	 gut	
Beurteilung «schlecht»	3,7	α = 3,4	p_y = 7,1
Beurteilung «gut»	β = 1,3	91,6	q_y = 92,9
Alle Werte in %	p_x = 5,0	q_x = 95,0	100,0

Selbst bei dem im Vergleich zur Fertigungsvarianz σ_x^2 «sehr genau» arbeitenden Auslesegerät des Beispiels 1 mit $\sigma_\varepsilon^2 = \sigma_x^2/100$ beträgt der falsch eingeordnete Anteil etwa 0,83%, also nahezu 1% der Fertigung. – Im Beispiel 2 würde man der Fertigung auf Grund der Messungen anstelle des realen Schlechtanteils $p_x = 5\%$ den fiktiven «Schlecht»anteil $p_y = 7{,}1\%$ zuordnen. Der vom Auslesegerät als «gut» bezeichnete Anteil $q_y = 92{,}9\%$ enthält noch 1,3% schlechte Teile.

5. Schlussbemerkungen

Zwei Toleranzgrenzen, $T_1 < \mu$ und $T_2 > \mu$ bzw. $\tau_1 < 0$ und $\tau_2 > 0$. Da die «Gewichtsfunktionen» $\Phi[(\tau - u)/\lambda]$ in Gleichung (3.5) bzw. $1 - \Phi[(\tau - u)/\lambda]$ in Gleichung (3.6) sogar für $\lambda = 1/2$ in der Umgebung von τ_1 bzw. τ_2 rasch gegen 0 streben, Fig. 3, so kann man die Auslesevorgänge an den beiden Grenzen *getrennt voneinander behandeln,* wenn τ_1 und τ_2 «genügend weit» auseinanderliegen. In beiden Fällen verwendet man die Formeln (4.11) und (4.12), wobei man einmal $\tau = \tau_1 < 0$ und einmal $\tau = \tau_2 > 0$ einsetzt. Im Sonderfall von symmetrisch zu μ gelegenen Grenzen mit $T_1 + T_2 = 2\mu$ bzw. $\tau_1 + \tau_2 = 0$ braucht man die aus (4.11) und (4.12) gefundenen Fehlausträge nur zu verdoppeln. Im Beispiel 2 des Abschnitts 4 würde man dann bei einer Fertigung mit dem Schlechtanteil $p_{x1} + p_{x2} = 10\%$ etwa 9,4% (!) der Fertigung falsch einstufen.

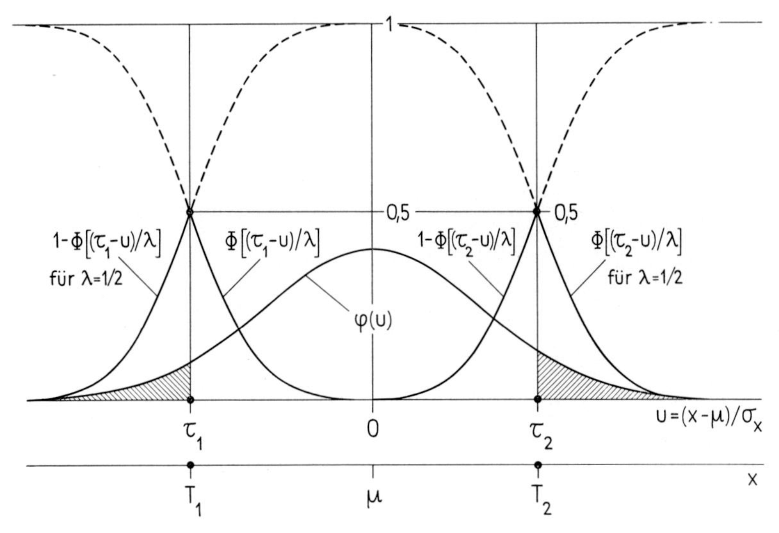

Fig. 3
Die Gewichtsfunktionen $\Phi[\;]$ und $1 - \Phi[\;]$ streben in der Umgebung von τ_1 bzw. τ_2 rasch gegen 0.

Kostenbetrachtung. Ist mit der falschen Einstufung eines guten bzw. schlechten Teils der Verlust A bzw. B verbunden und sind die Auslesekosten je Teil gleich C, so sind die Gesamtkosten des Auslesevorgangs (bei gegebener fester Toleranzgrenze τ) nur abhängig von λ, das heisst von der Güte des Messgeräts,

$$K(\lambda) = \underbrace{A\,\alpha(\lambda) + B\,\beta(\lambda)}_{K_{\mathrm{I}}} + \underbrace{C(\lambda)}_{K_{\mathrm{II}}}\,. \tag{5.1}$$

Nach (4.9) und (4.10) ist der mit Fehlentscheidungen verbundene Kostenanteil K_{I} eine mit λ monoton wachsende Funktion. Der Kostenanteil K_{II} ist eine mit

λ fallende Funktion, da Präzisionsgeräte mit «kleinem» $\lambda = \sigma_\varepsilon/\sigma_x$ bei der Beschaffung teurer, in der Pflege schwieriger und bei der Handhabung zeitaufwendiger sind als robuste Messgeräte mit «grossem» λ. Man findet demnach den in Fig. 4 skizzierten Verlauf der Kostenanteile $K_\mathrm{I}(\lambda)$ und $K_\mathrm{II}(\lambda)$ in Abhängigkeit von λ. Daraus kann man das kostengünstigste Ausleseverfahren (bzw. das zugehörige Gerät) mit $\lambda = \lambda_*$ bestimmen, welches die Gesamtkosten K zu K_* minimiert.

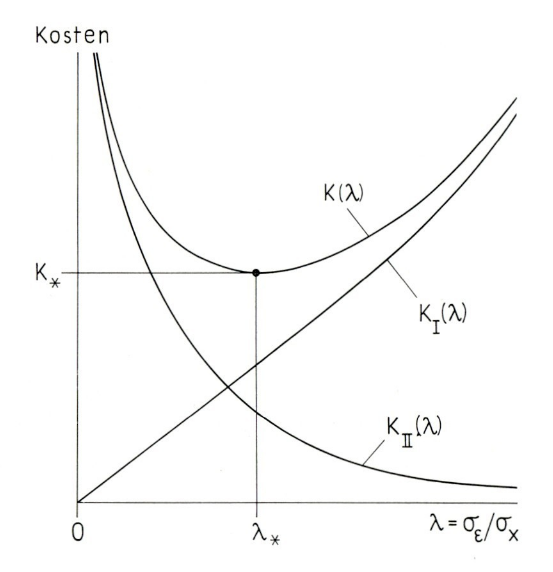

Fig. 4
Zur Bestimmung des kostengünstigsten Ausleseverfahrens (bzw. des zugehörigen Messgeräts) mit $\sigma_\varepsilon/\sigma_x = \lambda_*$. $K_\mathrm{I}(\lambda)$ sind die Kosten von Fehlentscheidungen; $K_\mathrm{II}(\lambda)$ sind die «reinen» Auslesekosten.

Schliesslich kann man die Toleranzgrenze τ' für die Messwerte y (die *Trenngrösse der Entscheidung*) gegen die Toleranzgrenze τ der gefertigten x-Werte (die *Trenngrösse der Beschaffenheit*) nach oben oder unten verschieben,

$$\tau' = \tau + \delta\,. \tag{5.2}$$

Ähnliche Überlegungen wie in den Abschnitten 2 bis 4 geben dann die Fehlausträge α und β in Abhängigkeit von $(\tau; \lambda; \delta)$. Bei fester Toleranzgrenze τ lässt sich dann die Kostenfunktion

$$K(\lambda; \delta) = A\,\alpha(\lambda; \delta) + B\,\beta(\lambda; \delta) + C(\lambda) \tag{5.3}$$

bezüglich λ und δ minimieren. Da $\partial\alpha/\partial\delta > 0$, $\partial\beta/\partial\delta < 0$ und $\partial C/\partial\delta = 0$ ist, so hängt die kostengünstigste Verschiebung $\delta = \delta_*$ dabei nur vom Verhältnis A/B, aber nicht von C ab. Mit diesen kurzen Hinweisen mag es an dieser Stelle sein Bewenden haben.

Adresse des Autors:
Institut für Statistik und Wirtschaftsmathematik, Rheinisch-Westphälische Technische Hochschule Aachen, D–51 Aachen, Pontstrasse 51.

C. Auer

C. Auer

Verfahren und Erfahrungen bei Insektenpopulationsschätzungen

1. Vorder- und Hintergründe

Das nachfolgende Beispiel der Forschungen über den *grauen Lärchenwickler*[1]) (Zeiraphera diniana, Gn.) steht in mancher Hinsicht stellvertretend für unzählige Fälle, wo der Jubilar A. Linder der Forschung und der Praxis ganz entscheidende Hilfen gab. Der fachliche Kontakt dauerte volle 25 Jahre. Kaum verwunderlich, dass in dieser Zeit auch ein tiefes menschliches Vertrauensverhältnis entstand. Reichlich verdient ist der bescheidene Dank mit dieser Veröffentlichung der Erlebniszusammenfassung.

Der graue Lärchenwickler ist ein unauffälliger Kleinschmetterling. Niemand würde ihm Beachtung schenken, wenn er nicht der grösste Schädling unserer Alpen-Lärchenwälder wäre. Seine kaum millimetergrossen Eier überwintern gut versteckt auf den Bäumen. Ungefähr gleichzeitig mit dem Austreiben der Lärchennadeln im Frühling schlüpfen kleinste Räupchen aus diesen Eiern, wandern über die Zweige, verstecken sich in den Lärchennadeln und ernähren sich auf diesem frischen Futter. Nach etwa 6 bis 7 Wochen sind die Raupen ausgewachsen, etwa 15 mm lang und 2 mm dick. Sie spinnen sich auf den Boden ab und verpuppen sich in der Streue-Mineralboden-Berührungszone. Nach 2 bis 3 Wochen schlüpfen die Falter, männliche und weibliche etwa zu gleichen Anteilen. Nach der Kopulation geht das Männchen verloren. Das Weibchen legt 100 bis 250 Eier und stirbt dann auch ab. Die neue Lärchenwicklergeneration hat begonnen. Unterdessen ist es Herbst geworden. Bis hierher ist an dieser Entwicklung gar nichts Aussergewöhnliches zu erkennen. In längerer Sicht, etwa alle 7 bis 10 Jahre, treten aber plötzlich so ungeheure Raupenmengen auf, dass sie ganze Bäume, Bestände, ja grosse Wälder entnadeln, mitten im Sommer. Nach 1 bis 3 Jahren verschwindet diese unliebsame Erscheinung wieder aus den Tälern, nie überall gleichzeitig. Solche Schaden-

[1]) Die Forschungen stehen unter Leitung des Entomologischen Institutes der ETH Zürich. Die statistischen Arbeiten werden vom Schweizerischen Nationalfonds für die wissenschaftliche Forschung getragen.

zeiten kennt man seit mindestens 100 Jahren aus allen Alpentälern. Was in der Zeitspanne dazwischen passiert, war bis 1949 aber noch ganz unbekannt.

So bescheiden waren die Kenntnisse, als sich der kantonale Forstdienst den lokalen Bestrebungen zur Verhinderung dieser Schäden helfend anschloss. Eine solche Aufgabe musste zunächst die Kenntnislücken in den Zwischenperioden schliessen, das Spiel der natürlichen Kräfte und Gegenkräfte erkennen, mit dem klar gestellten *praktischen Fernziel*, daraus naturangepasste Möglichkeiten zur Verhinderung dieser periodischen Massenvermehrungen zu entwickeln. Damit waren durch die Untersuchungen zum vornherein drei Grundsätze zu erfüllen:

- Übergang von den zeitlich zwischen den Schadenjahren aussetzenden Beobachtungen zu ihrer ununterbrochenen Fortsetzung.
- Statt der räumlichen Punktbeobachtungen musste ein sehr grosses Gebiet als Basis gewählt werden, um Falterflug- oder Ausweichbewegungen noch zu erkennen oder mindestens als Störeinflüsse auszuschalten.
- Für diesen zeitlich und räumlich ungewohnt grösseren Rahmen musste eine Methode zur *quantitativen Schätzung der natürlichen Populationen* entwickelt werden.

Daher und bereits in diesem frühen Zeitpunkt der Aufgabenstellung und -festlegung setzte die aktive Mitarbeit des mathematischen Statistikers A. Linder ein. Kurz vorher hatte er am Indian Statistical Institute, New Delhi (Mahalanobis), Bekanntschaft gemacht mit modernen Stichprobenverfahren und ihrer praktischen Durchführung. Die mengenmässige Erfassung von Insektenpopulationen grosser Waldgebiete, wie sie unsere Aufgabe erforderte, liess sich praktisch überhaupt nur mittels solcher Verfahren erfüllen und war für A. Linder daher ein geradezu gewünschtes Beispiel auf neuem Gebiet. So entstand die aussergewöhnlich fruchtbare Einflussnahme des Statistikers auf Form und Art der Forschungsarbeiten beim Lärchenwickler. Erst 1952, als sich die neue quantitative Schätztechnik bereits drei Jahre lang voll bewährt hatte, schaltete sich das Entomologische Institut der ETH Zürich unter P. Bovey als Fachwissenschaft helfend ein und übernahm die Leitung. Für den Forstdienst allein war diese Aufgabe zu gross geworden. Die bereits geschaffenen methodischen Verfahren bildeten aber nach wie vor die Grundlage aller Arbeiten und beeinflussten sie oft bis in kleine Einzelheiten.

2. Der Stichprobenplan und erste Erfahrungen

Forstamtliche Aufzeichnungen seit etwa 1830 ergaben für das Oberengadin die häufigsten und regelmässigsten Auftreten der LW[1])-Schäden innerhalb des Kantonsgebietes. Daneben sprachen noch viele andere Gründe für die Wahl dieses Gebietes als *Basis* für unsere Untersuchung. Das Gebiet zwischen Malojapass und dem Weiler Brail ist rund 35 km lang, hat eine Gesamtober-

[1]) LW = Abkürzung für grauen Lärchenwickler.

fläche von etwa 120 km^2 und trägt rund 6200 ha Wald sehr vielgestaltiger Art. Das forstliche Vollinventar zählt rund 613000 Lärchen über 16 cm BHD. Dieser Talabschnitt ist ausserdem zum grössten Teil durch hohe Gebirgszüge von andern Tälern getrennt.

Für diesen äusseren Rahmen musste ein Stichprobenverfahren entwickelt werden, das die mengenmässige Erfassung der LW-Populationen und ihrer Veränderungen von Jahr zu Jahr mit genügender Genauigkeit und bei tragbaren Kosten für eine lange Beobachtungsdauer ermöglichte. Das Schätzverfahren musste grundsätzlich die gleiche Wirksamkeit behalten bei hohen wie tiefen Populationen. Ferner mussten alle Einzelergebnisse eindeutig ihrem Fundort zugeordnet werden können. Lock-Fangmethoden irgendwelcher Art für die Falter schieden wir daher zum vornherein bewusst ganz aus. Die verfügbaren Geldmittel erlaubten nur eine einzige Aufnahme pro Jahr oder LW-Generation in diesem Rahmen. Arbeitstechnische Einsparungsmöglichkeiten spielten bei der Detailplanung eine bedeutende Rolle. Darum entschieden wir uns für die Erfassung des *Raupenstadiums* und beschränkten uns bewusst auf die Erfassung *relativer Raupendichten*, also der Anzahl Raupen auf einer bestimmten Lärchenzweigmenge. Zur Abschätzung aktiver Bekämpfungsmöglichkeiten interessierte uns auch die räumliche Verteilung der LW bei kleiner Dichte innerhalb des Untersuchungsgebietes. Wären in dieser Zeit eindeutige örtliche Verdichtungen (Herdgebiete) erkennbar geworden, so hätten diese der Bekämpfung besonders lohnende Ziele geboten.

Diese vielgestaltigen Gesichtspunkte erforderten einen Stichprobenplan, der das ganze einbezogene Waldgebiet lückenlos überdeckte. Dies erreichten wir mit drei Schichtungen (Talabschnitten, Talseiten und Höhenzonen) in zwei Stufen oder Schritten, wie in Tabelle 1 dargestellt.

Tabelle 1
Aufbau der Stichprobenpläne nach Zahl und Art der Schichten und Stufen.

Schicht Nr.	Element (und abgekürzte Bezeichnung)		Zahl der Elemente im Stpr. Plan 1949/57	ab 1958
1	Talabschnitt oder Gemeinden	TA	11	14
2	Talseiten	TS	2	2
	Teilgebiete (= TA × TS)	TG	22	28
3	Höhenzonen	HZ	2	3
	Kleingebiete (TG × HZ)[1]	KG	34	74
Stufe	**in der zufälligen Zuteilung der Probebäume (Pr. B)**			
1	auf die KG und direkt bis auf die forsteinrichtungstechnische Abteilung		in Gruppen (Gr) von 14–3 Pr. B gesamthaft	einzelbaumweise
2	Verteilung der Pr. B innerhalb der forstl. Abteilung		innerh. ökolog. einheitlicher Teile der Abt. mit ca. 500 Lärch.	einzeln innerhalb der ganzen Abt.

[1]) Die Höhenzonen waren nicht in allen Abteilungen in gleicher Zahl vorhanden.

Fig. 1 zeigt den Stichprobenplan des zweiten Zeitabschnittes auf das Untersuchungsgebiet übertragen. Es zeigt zugleich die Stichprobenergebnisse des Jahres 1970 in ihrer räumlichen Anordnung.

UNTERSUCHUNGSGEBIET OBERENGADIN

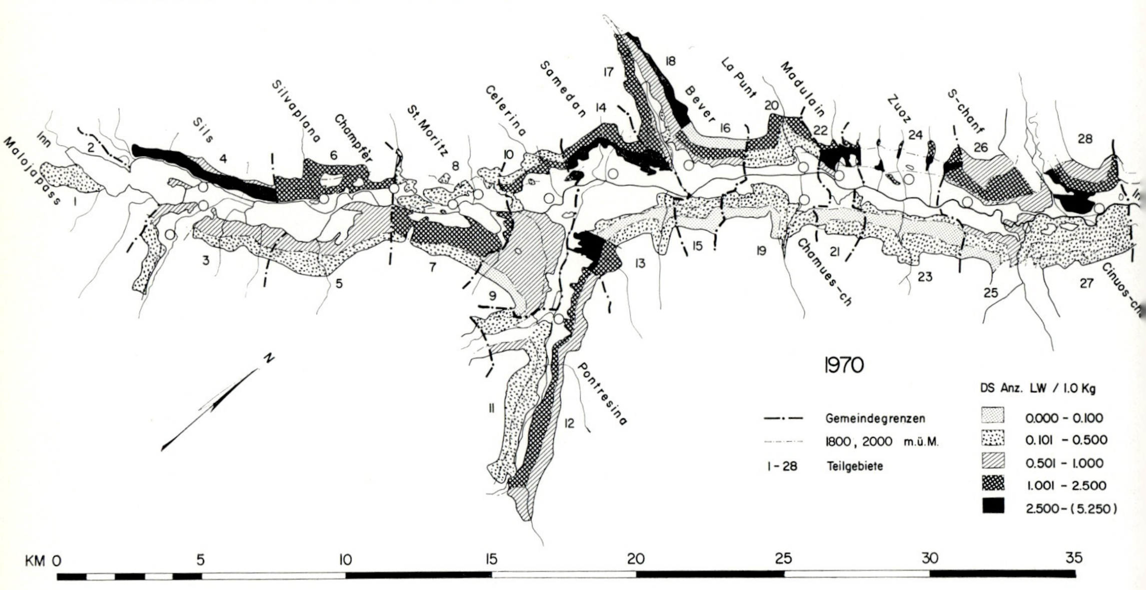

Fig. 1

Jedes Schichtelement konnte in freier Art nach Bedarf noch zusätzlich frei gewichtet werden. Dadurch blieb der Plan trotz festem Rahmen sehr anpassungsfähig. Davon haben wir in reichem Masse Gebrauch gemacht. A. Kaelin [1] hat diese Verfahren 1954 sehr eingehend beschrieben. Das Studium dieser Arbeit ist für jeden besonders Interessierten sehr lohnend.

Die Zeitspanne 1949 bis 1957 war für uns ein eigentlicher Lern- und Verbesserungsabschnitt. Dies sei allen jenen zum Trost gesagt, die vor ähnlichen Problemen stehen.

Den ersten Schritt anno 1949 mussten auch wir, trotz bester Vorausplanung, schliesslich doch aufs Geratewohl machen. Aber wir unternahmen ihn bewusst in vollem Massstab. Erst nach dem zweiten Arbeitsjahr hatten wir einigermassen genügende Unterlagen für den Beginn kritischer Überprüfungen. Es ging also alles nicht so schnell, wie man etwa annehmen könnte, zum Teil auch darum, weil wir die Vergleichbarkeit der Ergebnisse von Jahr zu Jahr sorgsam gewährleisten mussten.

3. Erfahrungen und Anpassungen

Was Kaelin [1] im Detail an Einzelbeispielen beschrieb, ist nachfolgend in wesentlichen Zügen für die ganze Zeit 1949–1957 zusammengefasst. Unser Bestreben ging vor allem dahin, den Aufwand für die Schätzungen zu verkleinern, ohne wesentliche Genauigkeits- oder Aussageverluste in Kauf nehmen zu müssen.

Für jedes Jahresergebnis konnten wir eine sogenannte hierarchische Streuungszerlegung berechnen.

Tabelle 2
Ergebnisse der hierarchischen Streuungszerlegungen über eine Massenwechselperiode.

Populations-dynamische Phase	Jahr	Durch-schnitt LW/1,0kg Zweige ($\bar{y}_w$)	Beobachtete Streuungsanteile der hierarchischen Stufen zwischen Bäumen innerhalb Gr	zwischen Gr innerhalb KG	zwischen KG innerhalb TA	zwischen TA innerhalb Tal
Progression	1950	0,082	0,015	0,86***	1,39*	12,58***
Progression	1951	0,445	0,32	0,61***	6,16***	14,79**
Progression	1952	4,174	12,10	60,70***	193,50***	442,80*
Progression	1953	68,820	1618,00	7296 ***	23462 ***	60801 **
Kulmination	1954	331,760	20092	93223 **	140622	2211582 ***
Regression	1955	126,540	5048	23279 ***	46828 **	129704 ***
Regression	1956	20,580	259	553 ***	1650 ***	5128 ***
Regression	1957	2,184	14,00	22 ***	20,20	109,70***

* Statistisch gesichert verschieden von der nächstunteren Stufe.

Diese Steurungsanteile ermöglichten uns jährlich eine Nachberechnung des günstigsten Stichprobenplanes. Daraus schöpften wir Hinweise für den Plan des folgenden Jahres.

Regelmässig wiederholt ergab diese kritische Überprüfung dabei ein besonders interessantes Ergebnis: Eine zufällige Verteilung der *einzelnen* Probebäume über das ganze Waldgebiet hätte wohl das genaueste Ergebnis gebracht. Bei der damaligen Zahl der Probebäume wäre ein solcher Plan aber nicht der wirtschaftlichste geblieben. Der Kostenanteil der langen und teuren Anmarschwege hätte zugenommen. Immerhin hatten diese Hinweise zur Folge, dass wir die Zahl der Bäume pro Stichprobengruppe von ursprünglich 14 sehr bald auf das Minimum von 3 herabsetzten. Daraus folgte bereits bis 1957 eine Halbierung der Probebaumzahl für das ganze Gebiet.

Tabelle 3
Vergleich der wirklich durchgeführten Stichprobenpläne mit nachberechneten Plänen und nur Einzelbaum-Stichproben (gekürzt).

Ergebnisse und Vergleiche	1949	1951	1953	1955	1957
1. Wirklicher Plan:					
Anz. Gruppen (Gr) n'	150	301	369	350	353
Anz. Pr. B/Gr n''	14	4	3	3	3
Total Anz. Pr. B	2100	1204	1107	1050	1059

Tabelle 3 (Fortsetzung)

2. Ergebn. wirkl. Plan:					
DS $\bar{y}_w$	0,1298	3,3337	131,7599	316,3527	5,4603
$SF_{\bar{y}}$	0,0118	0,3034	5,6205	10,2271	0,3571
100 $SF_{\bar{y}}/\bar{y}$ = % SF	9,09	9,10	4,27	3,23	6,54
3. $V\bar{y}_w$ wirklicher Plan	0,000140	0,092084	31,58970	104,5939	0,12751
$V^*\bar{y}$ berechneter Plan (gleiche Anz. Pr. B wie wirkl. Plan, prop. Stz auf Kleingeb. verteilt)	0,000087	0,043697	16,83506	63,4727	0,09859
$V^*\bar{y}/V\bar{y}_w$	0,563	0,474	0,533	0,608	0,773
Präzisionsgewinn mit Einzelstichpr. %	43,7	52,6	46,7	39,2	22,7
4. Anzahl nötige Einzelstpr. für gleiche $V\bar{y}_w$ wie wirkl. Plan	1182	554	590	637	819
5. $V\bar{y}_{400}$ für nur jährl. *400 Pr. B* prop. Stz auf TG verteilt	0,000462	0,131528	46,59104	166,6158	0,261018
$SF\bar{y}_{400}$	0,0215	0,3627	6,8258	12,9080	0,5109
100 $SF_{\bar{y}400}/y_{\bar{y}400}$ = % SF	16,55	10,88	5,13	4,08	9,36

Anfänglich benützten wir auch die Streuungen zwischen den Probebaumgruppen (Gr) innerhalb der Kleingebiete (KG) im Jahre i für die Gewichtung der Stichproben im Jahre $i + 1$. Dies war unzweckmässig. Die Veränderung der Streuungen konnte weder im Kleinen noch im Ganzen vorausgeschätzt werden. Als Gewichte benützten wir schliesslich nur mehr die Lärchenstammzahl pro KG.

Innert dieser Lernzeit gelang uns schrittweise auch eine zweckmässige Festlegung der *Stichprobengrösse* für den Probebaum, nämlich eine bestimmte Gewichtsmenge von Zweigen mit höchstens 5 bis 6 mm Durchmesser, zufällig der ganzen Baumkrone entnommen. 1,0 kg Zweige pro Baum scheint heute die untere verantwortbare Menge für einen ganzen Baum. Dies entspricht im Oberengadin annähernd 50 Laufmetern normal benadelter Zweige und einer Nadelfrischgewichtsmenge von etwa 440 g.

Damit war zugleich die Voraussetzung geschaffen, um den *Charakter der Verteilung der LW-Raupen* im Kleinen näher zu untersuchen. In der Regel entspricht die Häufigkeitsverteilung der LW-Raupen sowohl in den einzelnen Baumkronen wie in Beständen am besten der *negativen Binomialverteilung*. Nur bei höchster Raupendichte zeichnet sich eine Annäherung an die Normal- oder Gausssche Verteilung ab. Folgerichtig hätte man daher für alle mathematisch-statistischen Detailprüfungen mit logarithmisch transformierten Werten $(y' = \log(y + 1))$ rechnen müssen. In Hunderten von Fällen rechneten wir parallel mit den natürlichen und logarithmischen Werten. Es mag vor allem den Praktiker erfreuen, dass die Transformierung der Werte nur sehr selten ein Prüfresultat grundsätzlich änderte. Daher blieben wir bei der unmittelbar anschaulichen Ergebnisangabe in natürlichen Werten.

4. Der neue Stichprobenplan ab 1958

Ende 1957 überblickten wir zum ersten Mal einigermassen das Ausmass der LW-Dichteschwankungen, auch in Raum und Zeit. Die periodischen Auftreten der LW-Schäden hatten eine natürliche Erklärung gefunden, freilich in mancher Beziehung anders als erwartet. Sie waren gleichsam die sichtbaren Spitzen von Eisbergen im Meer (Fig. 2 als Beispiel).

QUANTITATIVER MASSENWECHSEL

DES LÄRCHENWICKLERS IM OBERENGADIN

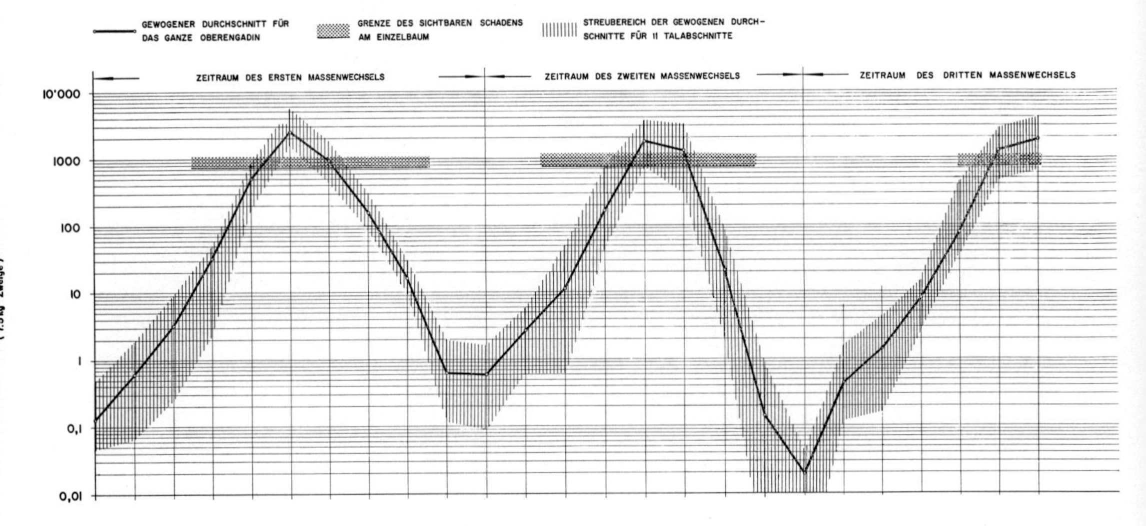

SCHADENAUSMASS UND SCHADENINTENSITÄT IN % DER TOTALEN WALDFLÄCHE (6281 ha)

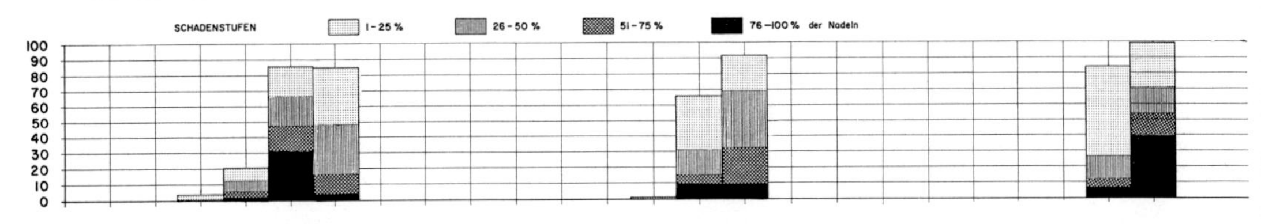

GEWOGENER DURCHSCHNITTLICHER NADELSCHADEN IN % FÜR DAS GANZE TAL

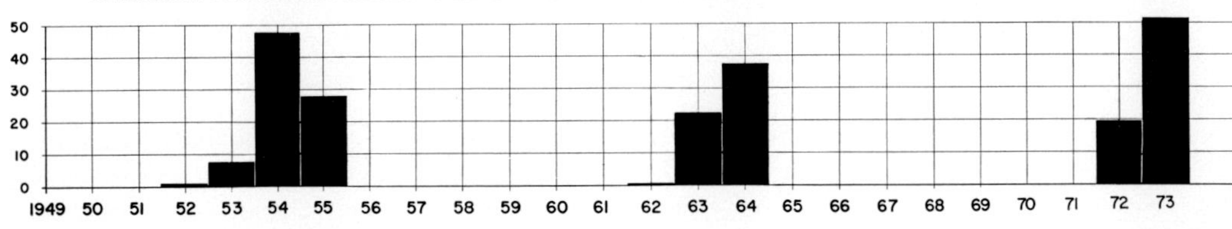

Fig. 2

Es war damit aber sehr bedeutsam geworden, die Untersuchungen fachlich vertieft fortzusetzen und insbesondere auch die Frage zu prüfen, ob der Massenbewegungsverlauf im Oberengadin ein Spezialfall sei oder sich im Optimumgebiet der Alpen auch anderswo und unabhängig wiederhole.

Für diese räumlich und zeitlich viel ausgedehntere Aufgabe musste ein einfacherer und doch genügend genauer Stichprobenplan entwickelt werden. Wir kannten nun die grossen jährlichen Sprünge in der Massenbewegung und durften daher die Ansprüche an die Genauigkeit der jährlichen Schätzungen verringern. Die Auskünfte über die räumliche Verteilung der LW-Dichten im Untersuchungsgebiet hatte inzwischen die mögliche praktische Bedeutung verloren. Es gibt keine sogenannten «Herdgebiete», wo ein «eiserner Bestand» die Zwischenschadenzeiten überdauert. An die Genauigkeit der *Nebeninformationen* (siehe Abschnitt 5) waren keine besonderen Ansprüche gestellt.

In den Mittelpunkt der Schätzungsziele war ganz unzweifelhaft die genügend genaue Erfassung der gewogenen durchschnittlichen Populationsdichte ($\bar{y}_w$) für das *ganze Tal* gerückt.

Damit war nun der Moment gekommen zu einer starken Herabsetzung der Probebaumzahl und zum Übergang zur zufälligen Einzelbaumverteilung. Diesen bedeutsamen Schritt haben wir durch Modellrechnungen anhand der wirklichen Zahlen aus der Untersuchung 1957 vorausgeprüft (Tabelle 4).

Tabelle 4
Ergebnisse von Modellberechnungen mit den Resultaten von 1957 (gekürzt).

Gebiet (Talabschnitt)	wirkl. Stichprobenplan 1957 353 Gr zu 3 Pr. B prop. Stz auf 34 Kleingebiete verteilt (DS 2,4 kg/Baum)			Modell a) 400 Pr. B prop. Stz auf 34 KG verteilt (DS 2,4 kg/Baum)			Modell b) gleich wie Mod. a), aber von jedem Baum fest 2,5 kg Zweige		
	$\bar{y}_w$	$SF_{\bar{y}}$	SF %	y_w	$SF_{\bar{y}}$	SF %	$\bar{y}_w$	$SF_{\bar{y}}$	SF%
Sils	2,766	0,519	19	2,966	0,696	23	4,465	1,084	25
Silvaplauna	5,152	0,764	15	5,768	1,191	21	5,391	1,172	22
St. Moritz	5,360	1,166	22	5,919	1,949	34	6,366	1,904	30
Celerina	5,236	0,709	14	7,270	1,490	20	7,896	1,641	26
Pontresina	5,627	0,886	16	4,800	1,001	21	5,380	1,141	21
Samedan	7,213	1,364	19	5,751	1,036	18	5,905	1,162	20
Bever	12,731	3,082	24	10,552	3,352	32	8,894	2,802	32
La Punt-Cham.	4,687	1,008	22	4,889	1,567	32	5,213	1,867	36
Madulain	4,956	1,365	28	5,665	1,055	19	6,068	1,451	24
Zuoz	6,050	0,901	15	5,112	1,135	22	4,689	1,087	23
S-chanf	3,211	0,562	18	2,139	0,765	36	2,284	0,889	39
Total	*5,460*	0,357	6,54	*5,170*	0,426	8,24	*5,399*	0,441	8,16

Wir wählten schliesslich den Modellfall *b* mit 400 Probebäumen, fester Stichprobenzweigmasse pro Baum und etwas verfeinertem Rahmen (Tabelle 1). Das Probengewicht verkleinerten wir von anfänglich 2,5 kg auf schliesslich 1,0 kg pro Baum.

Dieser Stichprobenplan hat sich in der Praxis sehr gut bewährt. Mit gewissen Anpassungen an die vorhandenen forsteinrichtungstechnischen Grundlagen, an die Geländeverhältnisse oder an besondere Ziele haben wir ihn in fünf verschiedenen Alpentälern längs des ganzen Alpenbogens und in verschiedenen Grossversuchen mit Erfolg angewandt. Beim LW können wir heute für alle Zwecke wirksame Schätzverfahren wählen.

5. Einige besondere Erkenntnisse

Üblicherweise sind statistische Verfahren nur Hilfsmittel für besondere Zwecke. Im vorliegenden Fall brachten sie aber zusätzlich *tragende Ergebnisse,* die sonst unerkannt geblieben wären; dies hauptsächlich dank der Ausdehnung der Untersuchung und des inneren Aufbaues des Stichprobenplanes.

Tabelle 2 belegt für jede neue hierarchische Stufe des Stichprobenplanes, die geographisch einer Erweiterung des Untersuchungsgebietes entspricht, das Auftreten neuer Einflüsse. Sie äussern sich im Auftreten zusätzlicher, statistisch gesicherter Streuungskomponenten. Wir können daraus erstmalig den

Tabelle 5
Gesamtpopulationen im Oberengadin und Präzision der Schätzwerte auf 7,5 kg Zweigmasse pro Baum.

Natürliche Werte Jahr	Gewogener DS pro 1000 Bäume $\bar{y}_i$	Standardfehler $SF\bar{y}_i$	Progressions-Faktor $\bar{y}_i/\bar{y}_{(i-1)}$
1949	134	12	
1950	613	38	4,5
1951	3334	303	5,4
1952	31303	935	9,4
1953	515979	16863	16,5
1954	2488197	69555	4,8
1955	949058	31176	0,38
1956	159602	5407	0,17
1957	16843	1067	0,11
1958	635	108	0,04
1959	598	102	0,94
1960	2785	227	4,7
1961	12284	1010	4,4
1962	171585	9149	14,0
1963	1866128	56862	10,88
1964	1382039	44134	0,74
1965	23371	1369	0,017
1966	146	54	0,0062
1967	18	18	0,1254
1968	441	84	24,120
1969	1475	138	3,344
1970	8007	444	5,428
1971	79267	3036	9,900
1972	1304490	45229	16,457
1973	1872088	64752	1,435

Beweis erbringen, dass es niemals möglich wäre, aus wenigen, noch so genauen Kleinstandortuntersuchungen allein einen gültigen Schluss auf die Massenbewegungen grosser Gebiete zu ziehen. Kleinstandortergebnisse lassen sich überhaupt nicht verallgemeinern. Dagegen darf man rückwärts gefahrlos Massenbewegungen grosser Gebiete nach Belieben in örtliche Teilbewegungen auflösen. Ebenso kann man ohne Zerstörung des als Ganzes gewonnenen Bildes die Wirkung von einzelnen oder verbundenen Standorteinflüssen näher prüfen, wie wir dies für die sogenannten *Nebeninformationen* machten.

Aus den bisherigen Ergebnissen ergibt sich für das Oberengadin eine *ununterbrochene Wellenbewegung* der LW-Dichten und damit auch der absoluten Populationen. Das Ausmass der Schwankungen übertrifft alle früheren Vermutungen (1:20000 bis 1:100000). Erkenntnistheoretisch von besonderer Bedeutung ist aber der gelungene Nachweis, dass jeder einzelne Massenwechsel (von Tief zu Tief gerechnet) durchaus eigene Charakterzüge aufweist. Massenwechsel müssen daher heute als *Einheiten* einer bisher unbekannten *höheren Ordnung* angesehen werden. Sie unterliegen offenbar anderen, eigenen Gesetzmässigkeiten und stehen über der Generation oder Population eines einzelnen Jahres. Für den Aufbau künftiger Forschungsarbeiten entstehen daraus sehr weitreichende Folgen.

Die Populationsschätzungen haben aber ausserdem jährlich eine reiche Fülle von *Nebeninformationen* erbracht. Sie alle ergänzen unsere Kenntnis über das innere Wesen des vielerwähnten biologischen Gleichgewichtes.

Ohne näher auf diese äusserst vielschichtigen Ergebnisse einzutreten, sei hier nur eine stichwortartige Aufzählung gegeben. Mögen sie den einen oder anderen Forscher ebenfalls für die Anwendung der Stichprobenverfahren gewinnen:

- Auskünfte über Zustand der LW-Raupenpopulationen wie: Altersstruktur nach Larvenstadien, Anteile gesunder, kranker, toter, parasitierter Raupen.
- Fleckigkeit der LW-Dichte innerhalb des Untersuchungsgebietes (Beispiel Fig. 1).
- Einflüsse des Stichprobenstandortes wie Bestandeszusammensetzung, Höhenlage (m ü. M.), Esposition, Bodenvegetation usw.
- Einflüsse der Probebaumeigenschaften wie ökologischer Standort, soziologische Stellung, Baumhöhe, BHD, Kronenausbildung.
- Einflüsse der Ameisen als Räuber.
- Zusammenhang LW-Dichte und LW-Schaden.
- Häufigkeitsverteilungscharakter für den LW.
- Populationsbewegungen von Begleitarten und Begleitartengruppen (Beispiel Fig. 3).

Alle diese Randbedingungen unterliegen in bestimmter, eigener Art ebenfalls Veränderungen innerhalb eines LW-Massenwechselraumes, zum Teil in sehr deutlich wiederholter Art. Auch solche Hinweise für das «Ganze» waren nur durch das angewandte Schätzverfahren erkennbar geworden.

Die massgebende Mitarbeit des Biometrikers schon bei den Planungsanfängen hat sich also äusserst fruchtbar ausgewirkt. Für die quantitative *Feldarbeit* der Entomologen ist eine neue Möglichkeit eröffnet worden, und zwar in einem Rahmen, der praktischen Bedürfnissen entspricht.

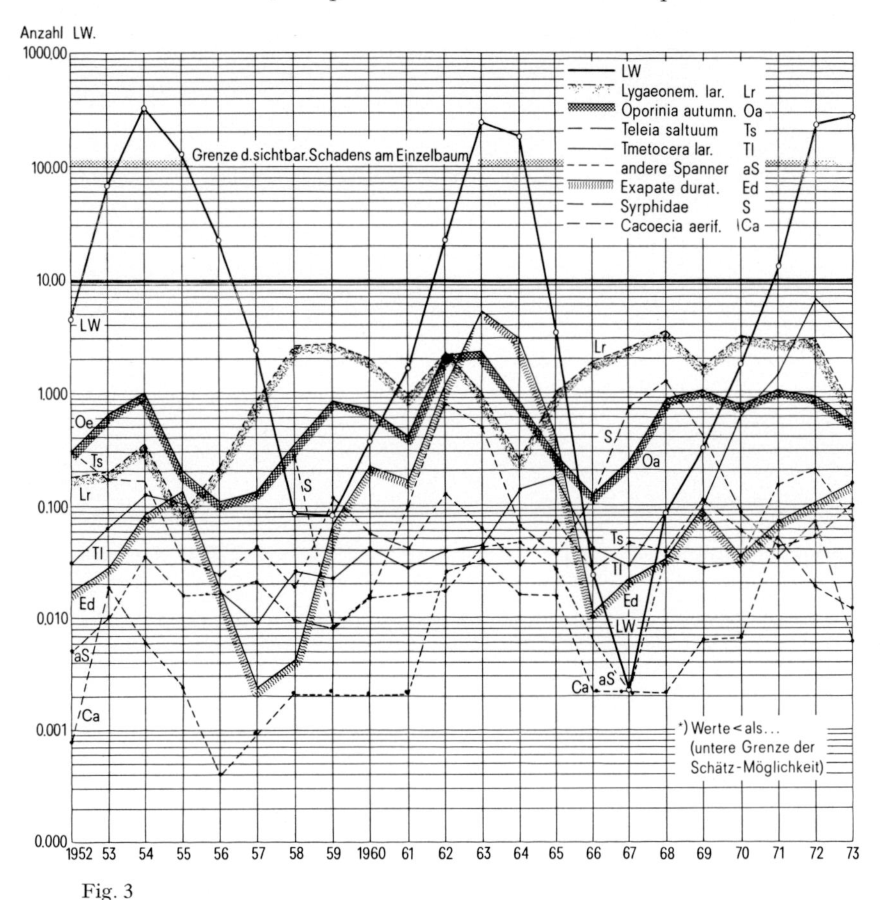

Fig. 3

Literaturhinweis

[1] KAELIN, A., und AUER, C. (1954): Statistische Methoden zur Untersuchung von Insektenpopulationen (dargestellt am Beispiel des grauen Lärchenwicklers). Z. f. angew. Ent. Bd. *36*, Hefte 3 u. 4, 80 S.
(Die inzwischen erschienenen, über 50 Facharbeiten der Teamkollegen sind hier absichtlich nicht erwähnt, weil der Interessent sie jederzeit über das Entomologische Institut der ETH Zürich erfahren kann.)

Adresse des Autors:
C. Auer, Dr. sc. tech. Dipl. Forsting. ETH/SIA, Arlibonstrasse 9, CH-7000 Chur.

3. Estimation

H. Berchtold und Th. Marthaler

Th. Marthaler

Vertrauensbereiche für den Quotienten von Mittelwerten zweier Normalverteilungen, wenn die Stichproben klein sind

1. Einleitung

Hat man zwei unabhängige Stichproben vom Umfang n_1 bzw. n_2 aus normalverteilten Grundgesamtheiten mit Mittelwerten μ_1 bzw. μ_2 und unbekannten Varianzen σ_1^2 bzw. σ_2^2, so interessieren üblicherweise Vertrauensbereiche für die Mittelwerte oder für die Differenz der Mittelwerte. Auf der anderen Seite werden viele Veränderungen nicht durch $\bar{x}_1 - \bar{x}_2$, dem Unterschied der Stichprobenmittelwerte, gemessen, sondern als Faktoren (zum Beispiel $\bar{x}_1$ ist das 1,4fache von $\bar{x}_2$) oder als prozentuale Erhöhung ($\bar{x}_1$ ist um 40% erhöht gegenüber $\bar{x}_2$). Verwendet der Benützer statistischer Methoden diese auf Quotienten oder Prozenten beruhende Ausdrucksweise, so sollte auch der Vertrauensbereich sich dieser Ausdrucksweise anpassen.

Dieses Problem ist jedoch im Hintergrund geblieben. Für grosse Stichproben finden sich vor allem in der zahnmedizinischen Literatur Beispiele.

Methoden zur Berechnung von Vertrauensbereichen für $\lambda = \mu_1/\mu_2$ stützen sich einerseits auf die Wahrscheinlichkeitsverteilung von $\bar{x}_1/\bar{x}_2$, (Geary [8]), andererseits auf diejenige von $\bar{x}_1 - \lambda\,\bar{x}_2$ (Bliss [2], Fieller [7]). Dubey [6] hat eine Zusammenstellung einiger Methoden zur Berechnung von Vertrauensintervallen gegeben, wenn ungleiche Varianzen vorliegen. Dasselbe Problem wurde weiter behandelt durch Chakravarti [3]. Sind über die Grundgesamtheiten schon vorgängige Informationen bekannt (zum Beispiel $\mu_1 < \mu_2$), so kann man das Vertrauensintervall unter Anwendung der Bayesschen Methode verkürzen. Das wurde von Abrams et al. [1] angewendet und diskutiert, jedoch nur für grosse Stichproben.

In der vorliegenden Arbeit wird die letztere Methode diskutiert, vor allem auch im Hinblick auf die Anwendung bei kleinen Stichproben; die auf Geary [8] zurückgehende Methode ist für kleine Stichproben ungeeignet. Ihre zusätzliche Verbesserung durch die Anwendung der Cornish-Fisher-Technik (Cornish et al. [6]) ist schon allein wegen des Rechenaufwandes nicht lohnenswert (siehe Dubey [6]).

2. Formeln

1. Fall: $\sigma_1 = \sigma_2$.

Sei $F_\alpha(i, j)$ das $(1-\alpha)$-Quantil der F-Verteilung mit (i, j) als Freiheitsgraden und $t_\alpha^2(j) = F_\alpha(1, j)$.

Ein Vertrauensbereich für $\lambda = \mu_1/\mu_2$ sei gegeben durch

$$\lambda^2(\bar{x}_2^2 - T^2 v_2^2) - 2\lambda \bar{x}_1 \bar{x}_2 + \bar{x}_1^2 - T^2 v_1^2 \leq 0, \tag{1}$$

wo $\bar{x}_i$ den Stichprobenmittelwert und v_i den Standardfehler der i-ten Stichprobe bedeuten. Es wird eine gemeinsame Standardabweichung s geschätzt, also gilt $v_i^2 = s^2/n_i$. T bedeute einen noch zu bestimmenden Faktor.

Ein exakter $(1-\alpha)$-Vertrauensbereich ergibt sich für $T = t_\alpha(n_1 + n_2 - 2)$. Dieser findet sich zum ersten mal bei Bliss [2], später verallgemeinert Fieller [7] diese Formel für abhängige Stichproben.

Sei $t_i = |\bar{x}_i|/v_i$ die Testgrösse zur Prüfung von $\mu_i = 0$ gegen $\mu_i \neq 0$ und $F = (t_1^2 + t_2^2)/2$ die Testgrösse zur Prüfung von $(\mu_1, \mu_2) = (0, 0)$ gegen $(\mu_1, \mu_2) \neq (0, 0)$.

Folgende drei Situationen sind möglich:

a) $t_2 > T$

Setzt man in (1) das Gleichheitszeichen, so besitzt diese quadratische Gleichung zwei Lösungen, die Vertrauensgrenzen von λ,

$$\lambda_{12} = \frac{\bar{x}_1 \bar{x}_2 \mp T\sqrt{\bar{x}_1^2 v_2^2 + \bar{x}_2^2 v_1^2 - T^2 v_1^2 v_2^2}}{\bar{x}_2^2 - T^2 v_2^2}, \quad \lambda_1 < \lambda_2 \tag{2}$$

oder

$$\lambda_{12} = \frac{v_1}{v_2} \frac{t_1 t_2 \mp T\sqrt{2F - T^2}}{t_2^2 - T^2}, \quad \lambda_1 < \lambda_2. \tag{3}$$

Der Vertrauensbereich ist das Intervall $\lambda_1 \leq \lambda \leq \lambda_2$.

b) $t_2 < T, F > T^2/2$

Auch hier berechnen sich die Vertrauensgrenzen gemäss (2), aber der Vertrauensbereich besteht aus

$\lambda \leq \lambda_1$ und $\lambda \geq \lambda_2$.

c) $t_2 < T, F < T^2/2$

Da ist die Ungleichung (1) für jedes λ richtig. Der Vertrauensbereich ist also $-\infty \leq \lambda \leq \infty$, was für den Benützer völlig nichtssagend ist.

Scheffé [10] nennt einen solchen Vertrauensbereich uneigentlich (improper), da der triviale Fall c) mit positiver Wahrscheinlichkeit eintreten kann. Um dies zu vermeiden, schlägt er folgende Formulierung des Problems vor:

a) Ist $F \leq F_\alpha\,(2, n_1 + n_2 - 2)$ und $\mu_2 \neq 0$, so macht man folgende Aussage: (μ_1, μ_2) liegt nicht ausserhalb der $(1-\alpha)$-Vertrauensellipse mit Mittelpunkt $(0, 0)$. Ein Vertrauensbereich für λ kann nicht bestimmt werden, da die Information über die Parameter ungenügend ist.

b) Ist $F > F_\alpha\,(2, n_1 + n_2 - 2)$ und $\mu_2 \neq 0$, so macht man folgende Aussage: λ liegt innerhalb [Fall a)] oder ausserhalb [Fall b)] der Vertrauensgrenzen, die durch (3) gegeben sind mit

$$T = t_\alpha(n_1 + n_2 - 2) + [F_\alpha(2, n_1 + n_2 - 2)/F]^r$$

$$\times\,[\sqrt{2\,F_\alpha(2, n_1 + n_2 - 2)} - t_\alpha(n_1 + n_2 - 2)]\,, \quad 0 \leq r \leq 1{,}35\,.$$

Dann ist die Wahrscheinlichkeit, eine richtige Aussage zu machen, $\geq 1 - \alpha$.

Bemerkungen:

a) Für den praktischen Wert $r = 1$ wird die Irrtumswahrscheinlichkeit etwas zu klein, für $r = 0$, das heisst $T^2 = 2\,F_\alpha\,(2, n_1 + n_2 - 2)$, wird sie erheblich zu klein. Ganz geringfügig zu gross werden kann sie für $r = 3/2$, was praktisch den besten Wert darstellt.

b) $t_2 = T$ oder $F = T^2/2$ können vorkommen, aber ihre Wahrscheinlichkeit ist Null; $t_2 > T$ und $F < T^2/2$ ist unmöglich wegen $t_2^2/2 \leq (t_1^2 + t_2^2)/2 = F < T^2/2 < t_2^2/2$.

2. Fall: $\underline{\sigma_1 \neq \sigma_2}$ (das heisst $v_i^2 = s_i^2/n_i$)

Vertrauensbereiche, deren wahre Irrtumswahrscheinlichkeiten exakt gleich der vorgegebenen sind, gibt es in diesem Fall nicht (Behrens-Fisher-Problem). Bekannte Approximationen sind folgende:

a) Der Vertrauensbereich ist durch (1) gegeben mit $T = t_\alpha(f)$ und

$$f = \frac{(v_1^2 + v_2^2)^2}{v_1^4/(n_1 - 1) + v_2^4/(n_2 - 1)}$$

gemäss einer Formel von Satterthwaite [9].

b) Der Vertrauensbereich ist durch (1) gegeben mit

$$T = \frac{v_1^2\, t_\alpha(n_1 - 1) + v_2^2\, t_\alpha(n_2 - 1)}{v_1^2 + v_2^2}$$

nach einer Approximation von Cochran [4].

c) Der Vertrauensbereich ist gegeben durch

$$\lambda^2\left(\bar{x}_2^2 - t_\alpha^2(n_2 - 1) \cdot v_2^2\right) - 2\,\lambda\,\bar{x}_1\,\bar{x}_2 + \left(\bar{x}_1^2 - t_\alpha^2(n_1 - 1) \cdot v_1^2\right) \leq 0\,, \qquad (4)$$

wobei man wieder die drei möglichen Fälle unterscheiden kann. Falls Vertrauensgrenzen existieren, berechnen sie sich zu

$$\lambda_{12} = \frac{\bar{x}_1\,\bar{x}_2 \mp \sqrt{t_\alpha^2(n_1 - 1) \cdot v_1^2\,\bar{x}_2^2 + t_\alpha^2(n_2 - 1)\, v_2^2\,\bar{x}_1^2 - t_\alpha^2(n_1 - 1) \cdot t_\alpha^2(n_2 - 1)\, v_1^2\, v_2^2}}{\bar{x}_2^2 - t_\alpha^2(n_2 - 1) \cdot v_2^2}$$

$$= \frac{v_1}{v_2}\; \frac{t_1\,t_2 \mp \sqrt{t_\alpha^2(n_1 - 1) \cdot t_2^2 + t_\alpha^2(n_2 - 1) \cdot t_1^2 - t_\alpha^2(n_1 - 1)\, t_\alpha^2(n_2 - 1)}}{t_2^2 - t_\alpha^2(n_2 - 1)}\,.$$

Der Vertrauensbereich (4) garantiert eine Vertrauenswahrscheinlichkeit $\geq 1 - \alpha$. Das wurde von Chakravarti [3] bewiesen. Er hat auch eine Publikation angekündigt, die das Vorgehen von Scheffé auf den Fall ungleicher Varianzen übertragen soll. Bisher ist uns aber noch nichts darüber bekannt.

Literatur

[1] Abrams, A. M., McClendon, B. J., and Horowitz, H. S. (1972): Confidence intervals of percentage reductions. Journal of Dental Research *51*, 492–497.

[2] Bliss, C. I. (1935): The comparison of dosage-mortality data. Ann. Appl. Biology *22*, 307–333.

[3] Chakravarti, I. M. (1971): Confidence set for the ratio of means of two normal distributions when the ratio of variances is unknown. Biometr. Z. *13*, 89–94.

[4] Cochran, W. G. (1964): Approximate significance levels of the Behrens-Fisher test. Biometrics *20*, 191–195.

[5] Cornish, E. A., and Fisher, R. A. (1937): Moments and cumulants in the specification of distributions. Rev. de l'Inst. Int. de Stat. *5*, 307–322.

[6] Dubey, S. D. (1966): On the determination of confidence limits of an index. Biometrics *22*, 603–609.

[7] Fieller, E. C. (1940): The biological standardization of insulin. Suppl. J. Roy. Statist. Soc. *7*, 1–64.

[8] Geary, R. C. (1930): The frequency distribution of the quotient of two normal variates. J. Roy. Statist. Soc. *93*, 442–446.

[9] SATTERTHWAITE, F. E. (1946): An approximate distribution of estimates of variance components. Biometrics Bull. *2*, 110–114.
[10] SCHEFFÉ, H. (1970): Multiple testing versus multiple estimation. Improper confidence sets. Estimation of directions and ratios. Ann. Math. Stat. *41*, 1–29.

Adressen der Autoren:
H. Berchtold, Th. Marthaler, Biostatistisches Zentrum der Medizinischen Fakultät der Universität Zürich, Plattenstrasse 54, 8032 Zürich.

P. Bauer, V. Scheiber
and F. X. Wohlzogen

F. X. Wohlzogen

Sequential Estimation of the Parameter π of a Binomial Distribution

Summary

A sequential procedure is proposed for the graphical estimation of the parameter π of a binomial distribution. It is based on the following argument: As soon as in the course of sequential sampling the two hypotheses, $H_{\pi_0+}: \pi > \pi_0 + d$ and $H_{\pi_0-}: \pi < \pi_0 - d$, $0 < d \leqslant \min(\pi_0, 1 - \pi_0)$, can both be rejected with an error probability $\leqslant \alpha$ in either case, an interval estimate $[\pi_0 - d, \pi_0 + d]$ is obtained, which covers the true parameter π with Prob $\geqslant 1 - 2\alpha$.

The construction of estimation curves is described and illustrated by examples for various values of d and α. These curves are compared with those found by Malý [4] which, for the same values of d and α, consistently require larger samples for any estimate of π. The use of the various decision boundaries in the graphical estimation of π is demonstrated by a biological example and their possible application to a certain type of clinical trials is discussed.

1. Introduction

Given a population of N units with R 'reacting' units or 'successes' and S 'silent' (= not reacting) units or 'failures' ($S = N - R$), let $\pi = R/N$. For an infinite population π is the parameter of a binomial distribution. In estimating π from a sample of size n, confidence intervals for π (Clopper and Pearson [3]) are obtained, the lengths of which vary with the observed maximum likelihood estimate $\hat{\pi} = r/n$, where r is the number of 'reactors' in the sample. There are practical situations, however, where the primary concern is not so much to arrive at a maximum likelihood estimate $\hat{\pi}$, but to obtain an interval estimate of π with given constant length, and any chosen confidence coefficient $1 - 2\alpha$, irrespective of the value of $\hat{\pi}$. Obviously this is impossible by the Clopper-Pearson method with fixed sample size techniques.

For example, for $n = 50$ an observed estimate $\hat{\pi} = 0.5$ leads to a 95 per cent confidence interval for π of [0.36, 0.64], whereas for $\hat{\pi} = 0.9$ the respective interval is [0.78, 0.97]. In the case of $\hat{\pi} = 0.5$ one would need a sample of size $n \sim 110$ for getting a confidence interval as small as in the case of $\hat{\pi} = 0.9$ with $n = 50$.

In this paper a sequential estimation procedure for π is introduced, which provides interval estimates of given constant length, independent of the value of the observed estimate for π.

2. Principle of the Procedure: The Method of Malý

Malý [4] approached the problem of obtaining interval estimates of constant length for any value of π, by applying k sequential probability ratio test (SPR-tests; Wald [6]) for testing k hypotheses about the unknown parameter π of a binomial distribution. One of the k hypotheses $H_{\pi_i}: \pi = \pi_i$ $\left(i = 1(1)\,k,\ d \leqslant \pi_i \leqslant 1 - d,\ 0 < d < 0.5\right)$ is to be accepted, as soon as the two auxiliary hypotheses, $H_{\pi_i+}: \pi = \pi_i + d$ and $H_{\pi_i-}: \pi = \pi_i - d$, when tested against H_{π_i}, can *both* be rejected with an error probability $\leqslant \alpha$ in either case. Hence the true parameter π is covered with probability $\geqslant 1 - 2\,\alpha$ by the interval

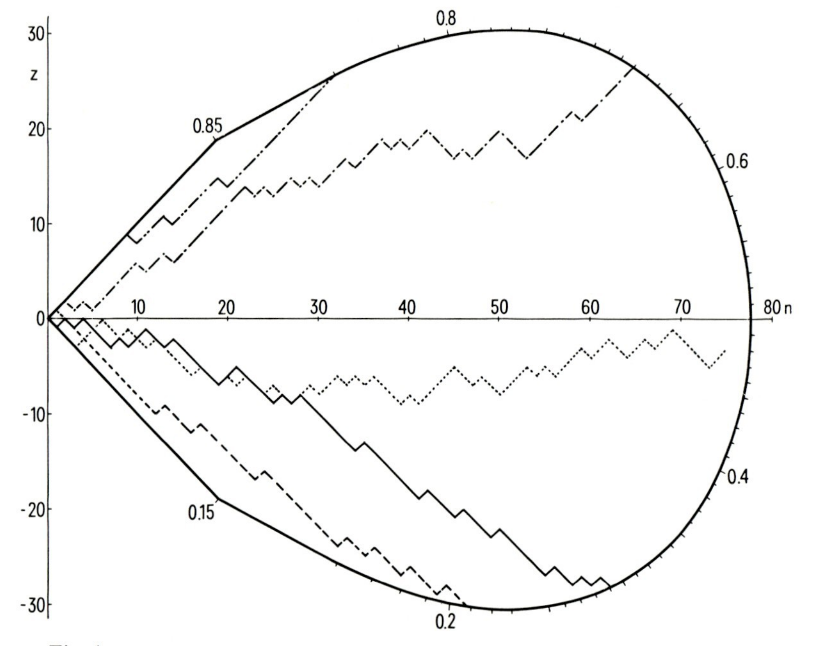

Fig. 1
Decision curve obtained by Malý's procedure, with $d = 0.15$ and $\alpha = 0.025$.
The scaling on the curve denotes the different estimates $\dot{\pi}$. The sequential sample paths stem from experiments with animals from a population with known distribution of the individual threshold doses (see text). They were given the effective doses ED 20 (– – –), ED 27 (———), ED 50 (· · ·), ED 73 (–·–·–), and ED 80 (–··–··).

$[\pi_i - d, \pi_i + d]$. Malý's original idea was, that for $k \to \infty$ a continuous estimation curve for all values of π $(d \leqslant \pi \leqslant 1 - d)$ results from the intercepts $\dot{\pi}$ of the upper (lower) decision lines in the SPR-tests for H_π versus $H_{\pi-}$ $(H_{\pi+})$. Plotting these intercepts $\dot{\pi}$ in a n, z-coordinate system $(z = 2r - n)$, and calculating sufficiently many values of $\dot{\pi}$, the estimation curve can be drawn (by connecting adjacent $\dot{\pi}$) and scaled (by $\dot{\pi}$) to any required degree of accuracy. It is pear-shaped and symmetrical with respect to $z = 0$, as can be seen from the example in Fig. 1.

The experimental results are plotted in the usual manner (e.g. Armitage [1]): If in the n-th sample unit a 'success' was observed, the value of $z(n)$ is calculated as $z(n) = z(n-1) + 1$, and for a 'failure' $z(n) = z(n-1) - 1$. Hence the 'sample path' moves in a zig-zag line, starting at the origin and moving one unit diagonally upwards for an observed success and one unit diagonally downwards for a failure. In Fig. 1 some examples (Wohlzogen et al. [8]) of such paths are shown, which will be discussed in Section 5.

Malý's graphs, of which the estimation curve in Fig. 1 is an example, have the following properties: As soon as a sample path reaches or crosses the curve the test is terminated and the intercept of the sample path with the curve yields not only a point estimate $\dot{\pi}$, but also an interval estimate $[\dot{\pi} - d, \dot{\pi} + d]$ for π, with a confidence coefficient $\geqslant 1 - 2\alpha$. Actually the confidence coefficient must be appreciably larger than $1 - 2\alpha$: According to the rules of SPR-tests the acceptance of H_π against $H_{\pi+}$ and $H_{\pi-}$ can occur already 'inside' the estimation curve, viz. whenever a sample path has reached or crossed the rejection lines for *both* hypotheses; $H_{\pi+}$ and $H_{\pi-}$, against H_π in either case. For this reason the confidence coefficients of the interval estimates $[\dot{\pi} - d, \dot{\pi} + d]$ obtained by Malý's method must be systematically higher than their nominal value $1 - 2\alpha$. This will be illustrated by an example in the Appendix.

3. The HT-Procedure, a Method Based on Non-sequential Hypothesis Testing

As explained in the preceeding section, using decision lines of the SPR-tests for the construction of estimation curves will always provide higher confidence coefficients for the estimates than actually desired. An intuitively obvious modification of Malý's method would be, to use the confidence limits according to fixed sample size theory in the construction of estimation curves, irrespective of the sequential nature of sampling. This procedure, based on non-sequential hypothesis testing (HT-procedure), will be outlined in this section.

The joint rejection of the two hypotheses, $H_{\pi_0+}: \pi > \pi_0 + d$ *and* $H_{\pi_0-}: \pi < \pi - d$, with an error probability $\leqslant \alpha$ in either case, will lead to an interval estimate $[\pi_0 - d, \pi_0 + d]$ for π with confidence coefficient $\geqslant 1 - 2\alpha$. The problem is to find the smallest value of n for any π_0, d and α fixed, for which the hypotheses H_{π_0+} and H_{π_0-} can *both* be rejected.

For a sample of size $n(n \geqslant n_{min}(d, \alpha))$[1]) there exists a value $r_l = r_l(\pi_0 + d, n, \alpha)$ so that H_{π_0+} may be rejected with an error probability $\leqslant \alpha$, if the number of observed successes $r \leqslant r_l$. Analogously, for a sample of size n $(n \geqslant n_{min})$[1]) there exists a value $r_u = r_u(\pi_0 - d, n, \alpha)$, so that H_{π_0-} may be rejected with an error probability $\leqslant \alpha$, if the number of observed successes $r \geqslant r_u$. The boundaries $r_l(r_u)$ can be derived as the largest (smallest) integers satisfying the inequalities (Clopper and Pearson [3]):

$$\sum_{x=0}^{r_l} \binom{n}{x} (\pi_0 + d)^x (1 - \pi_0 - d)^{n-x} \leq \alpha\,, \tag{1}$$

$$\sum_{x=r_u}^{n} \binom{n}{x} (\pi_0 - d)^x (1 - \pi_0 + d)^{n-x} \leq \alpha\,. \tag{2}$$

Due to the descreteness of x it is generally impossible to find integers r_l and/or r_u, for which the equality sign holds in (1) and (2). Real numbers, however, $r_l' = r_l'(\pi_0 + d, n, \alpha)$ and $r_u' = r_u'(\pi_0 - d, n, \alpha)$ may be interpolated which formally satisfy the equality sign. If, for increasing values of n, r_l' (r_u') are divided by n, plotted against n, and connected by straight lines, two polygons r_l'/n (r_u'/n) are obtained. Obviously these polygons must cross at a finite real $n'(\pi_0, d, \alpha) \geqslant n_{min}(d, \alpha)$: For values of $n < n'_{min}$, r_l'/n lies below r_u'/n and approaches $\pi_0 + d$ when $n \to \infty$, whereas the polygon r_u'/n, for $n \to \infty$, approaches $\pi_0 - d$. The integer n, $n - 1 < n' \leqslant n$, denotes the sample size when, for the first time, it may be possible to observe a value of r, so that the two hypotheses, $H_{\pi_{0+}}$ and $H_{\pi_{0-}}$ can both be rejected simultaneously with an error probability $\leqslant \alpha$ in either case, and the true parameter π is covered with probability $\geqslant 1 - 2\,\alpha$ by the interval $[\pi_0 - d, \pi_0 + d]$.

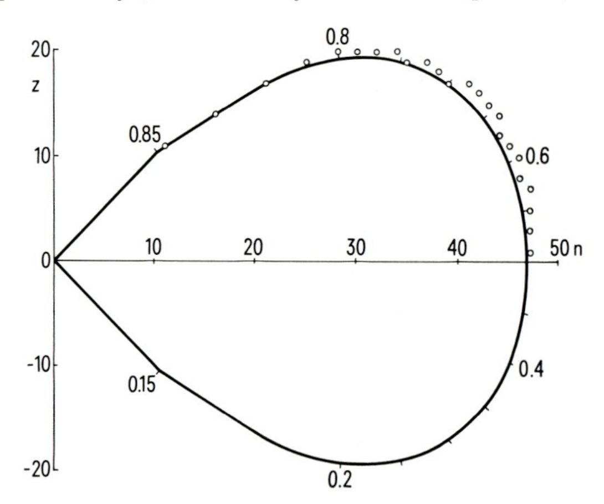

Fig. 2
Decision curve obtained by the HT-procedure, with $d = 0.15$ and $\alpha = 0.025$.
The scaling on the curve denotes the different estimates $\tilde{\pi}$. The actual end points of all possible sample paths, on or immediately outside the curve are shown for $z \geq 0$.

[1]) For the calculation of n_{min} see A 1.

In an analogous way as described in the previous section—calculating the intercepts $\tilde{\pi}$ of the two polygons, r_l'/n and r_u'/n, and the corresponding n' in the interval $[n-1, n]$ for sufficiently many values of π_0, and plotting $\tilde{\pi}$ in a n, z-coordinate system $(z = 2r - n)$—an estimation curve for π (HT-graph) can be drawn and scaled (by $\tilde{\pi}$) to any required degree of accuracy.

These HT-graphs resemble the curves obtained by Malý's method: They, too, are pear-shaped and symmetric to $z = 0$ (see Fig. 2). Their use is also analogous to Malý's graphs: If a sample path reaches or crosses the curve at the point denoted $\tilde{\pi}$, one gets an interval estimate $[\tilde{\pi} - d, \tilde{\pi} + d)$ for π with a confidence coefficient $\geqslant 1 - 2\alpha$. This will be shown in detail in A2 for the actual decision points in Fig. 2.

The values of $r_l'(\pi_0 + d, n, \alpha)$ and $r_u'(\pi_0 - d, n, \alpha)$ for $\pi_0 \neq 0.5$ are not symmetrical with respect to π_0 and the estimate $\tilde{\pi}$ is not identical with the maximum likelihood estimate $\hat{\pi}$. The more $\tilde{\pi}$ approaches d or $1 - d$, the more it differs from $\hat{\pi}$. If a sample path consists of observed successes (failures) only, the difference becomes as large as d. The HT-procedure, however, aimes primarily at interval estimates of given length $2d$ to be read directly from the graph, the confidence coefficient of which is at least $1 - 2\alpha$.

In Fig. 3 five HT-graphs are shown for the values $\alpha = 0.025$, $d = 0.1$, 0.15, 0.2; $d = 0.15$, $\alpha = 0.05$, 0.005. From these graphs it is apparent, that the boundaries will be affected more strongly by the choice of d than by the choice of α. It will be up to the experimenter to choose a suitable combination of d and α for his special problem.

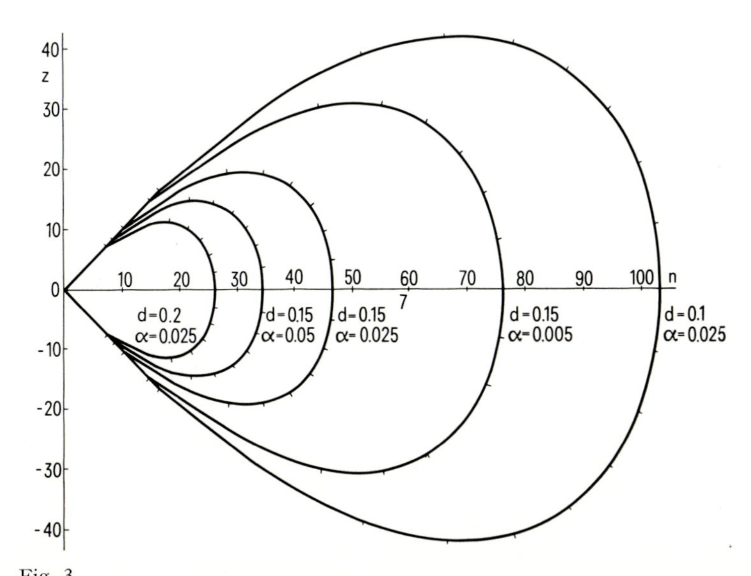

Fig. 3
Comparison of curves obtained by the HT-procedure for different combinations of d and α.
The scaling on the curves denotes the estimates $\tilde{\pi}$ in steps of 0.05 ($\tilde{\pi} = 0.5$ when $z = 0$).

4. Discussion

The HT-procedure, as outlined in Section 3, for obtaining interval estimates of given constant length $2d$ for π, independent of the value of π, with confidence coefficient $\geqslant 1 - 2\alpha$, is based on hypothesis testing according to fixed sample size theory. It is applied, however, to sampling of an essentially sequential nature.

On the other hand, Malý's method of constructing estimation curves is based on individual SPR-tests, but neglects the possibility of arriving at an interval estimate $[\dot{\pi} - d, \dot{\pi} + d]$ for π before a sample path has reached the estimation curve. That is why this method requires larger $n(d, \alpha)$ than the HT-procedure (see A1). From a comparison of Fig. 1 and Fig. 3 it can be seen, that Malý's graph ($d = 0.15$, $\alpha = 0.025$) corresponds approximately to the HT-graph ($d = 0.15$, $\alpha = 0.005$). In fact, calculation of the confidence coefficients provided by the graph of Fig. 1 always led to a value $\geqslant 0.99$.

The actual, numerically determined confidence coefficients in the HT-procedure are never smaller than $1 - 2\alpha$. For the extreme cases, sample paths with $z = n$ or $z = -n$, when only successes or failures are observed in a sample, one tail of the error probabilities for the estimates, $[1 - 2d, 1]$ and $[0, 2d]$ respectively, vanishes and the confidence coefficients become larger than $1 - \alpha$.

The apparent gain, when using the HT-procedure instead of Malý's method is shown for n_{min} and n_{max} in Table 1 (for the same combinations of d and α as in Fig. 3).

Table 1
Comparison of n_{min} and n_{max} in Malý's method and the HT-procedure for the decision curves with $\alpha = 0.025$, $d = 0.1, 0.15, 0.2$; $d = 0.15$, $\alpha = 0.05, 0.005$.

		$\alpha = 0.025$			$d = 0.15$	
		$d = 0.2$	$d = 0.15$	$d = 0.1$	$\alpha = 0.05$	$\alpha = 0.005$
n_{min}	Malý	12.7	18.9	31.1	15.2	27.3
	HT	7.2	10.3	16.5	8.4	14.9
n_{max}	Malý	42.0	77.7	179.5	62.4	112.3
	HT	26.4	47.0	103.5	34.8	76.5

It should be mentioned, that there exists a general inequality for sequential procedures (Wald [6], Lemma 10:6), which can be used for the construction of sequential estimation procedures. A graphical estimation method based on this inequality leads to exactly the same n_{min} as the HT-procedure, whereas the value for n_{max} (when $\pi = 0.5$) is considerably larger than in the HT-procedure and shows good agreement with n_{max} of Malý's method. Other principles, too, may govern the construction of estimation procedures, e.g. the fiducial argument. It will lead, when applied to the binomial distribution, to estimation curves different from the HT-procedure, generally requiring smaller samples.

5. An Example

In Fig. 1 the sample paths observed in an animal experiment with 5 doses of a biologically active substance (chorionic gonadotrophin) eliciting an all-or-none-response (spermiation) in male toads (Wohlzogen [7]) are shown (Bukovics et al. [2]). The dose-response relation in the animal population used was known from long term experience. The effective doses ED 20, ED 27, ED 50, ED 73, and ED 80 were given up to 75 animals each.

Hence in Fig. 1 the sample path for ED 50 ends at $n = 75$, before the estimation curve has actually been reached. Arguing that a sample path can only move diagonally upwards or downwards, the only possible estimates $\dot{\pi}$ from Malý's graph must lie between 0.47 and 0.50.

In Table 2 Maly's method and the HT-procedure, each for $d = 0.15$ and $\alpha = 0.025$, are compared when applied to the experiment described above.

Table 2
Sample sizes n needed and estimates obtained by Malý's method ($\dot{\pi}$) and the HT-procedure ($\tilde{\pi}$), when applied to the experiment shown in Fig. 1 ($d = 0.15$, $\alpha = 0.025$).

Dosis	π	Malý		HT	
		n	$\dot{\pi}$	n	$\tilde{\pi}$
ED 20	0.20	47	0.21	28	0.19
ED 27	0.27	63	0.29	39	0.30
ED 50	0.50	78	0.47–0.50	47	0.56
ED 73	0.73	65	0.69	37	0.73
ED 80	0.80	32	0.84	25	0.82

As can be seen from Table 2 the deviations of Malý's estimate $\dot{\pi}$ from the 'true' values are never larger than 0.04 requiring sample sizes appreciably larger than the HT-procedure. Nevertheless the estimates $\tilde{\pi}$ obtained by the HT-procedure are also well within the tolerated deviations $\pi \pm d$.

6. Application of the HT-Procedure to Medical Trials; a Treatment of Ties

A problem, frequently arising in clinical trials, is to decide, whether a treatment A is superior to another treatment B. A powerful technique in this situation is that of paired comparisons, with the decision based on the proportion of pairs with preference in one direction, 'A better than B', say. In many instances it will be advantageous to plan the investigation as a sequential trial (cf. Armitage [1]). There are practical situations, however, when the comparison of the two treatments should be more specified, i.e. an inference of the following form is desirable: The proportion of untied pairs, in which A gives a better result than B, lies between $\pi - d$ and $\pi + d$, with a confidence coefficient $\geqslant 1 - 2\alpha$. In these cases the HT-procedure might prove useful. The HT-graph would be used then, counting the pairs as units, and the result 'A better than B' in a pair as a success, say, and 'B better than A' as a failure. The number of pairs required in any test (d, α) will depend on the value of π and on the number of ties encountered. A pair in which the effects of A and B are equal is called a tie and does not count in the analysis, i.e. the sample path 'halts' whenever a tie is observed. Under certain conditions it is reasonable to substitute this conventional method of treating ties, by another strategy.

If it is a continuous variable, by which the effects of the two treatments A and B are judged, e.g. the depression of arterial blood pressure, the prolongation of clotting time, etc., theoretically no ties—exactly equal results of treatment A and B in a pair—should occur. Because of limited accuracy in measurements, nevertheless ties may and do appear in practical situations. The following procedure is proposed for treating such ties: For the first tie observed in the sequential sample it is randomly decided whether it is to be counted as a success or a failure. The following ties are always counted as events contrary to that ascribed to the preceding tie.

This procedure of treating ties in those situations, where the occurrence of a tie is merely due to the limited accuracy of individual measurements, has been tested by extensive computer simulations. These and the results obtained are described in A3.

Appendix

A1. *Calculation of n_{min} for the HT-Procedure*

The estimates $\tilde{\pi} = d$ $(\tilde{\pi} = 1 - d)$ are the lowest (highest) values of $\tilde{\pi}$ providing an interval estimate of length $2\,d$, $\tilde{\pi} - d \geqslant 0$, $\tilde{\pi} + d \leqslant 1$. The smallest value $n_{min} = n_{min}(d, \alpha)$ of $n'(\pi_0, d, \alpha)$, for which the joint rejection of $H_{\pi_{0-}}$ and $H_{\pi_{0+}}$ is possible, can therefore be calculated for $\pi_0 = d$ or $\pi_0 = 1 - d(|z'| = n')$:

$$\min_{\pi_0}\left(n'(\pi_0, d, \alpha)\right) = n'(d, d, \alpha) = n'(1 - d, d, \alpha) = n'_{min}(d, \alpha)\,, \tag{A1}$$

$$(1 - 2\,d)^{n'_{min}(d,\alpha)} = \alpha; \quad n'_{min}(d, \alpha) = \frac{\ln \alpha}{\ln(1 - 2\,d)}\,, \tag{A2}$$

$$0 < d < 0.5 - \alpha/2\,. \tag{A3}$$

For comparison $\dot{n}_{min}(d, \alpha)$ in Malý's procedure is

$$\dot{n}_{min}(d, \alpha) = \frac{\ln \dfrac{1 - \alpha}{\alpha}}{\ln \dfrac{(1 - d)}{(1 - 2\,d)}}\,. \tag{A4}$$

It can be shown that for any d, $0 < d < 0.5 - \alpha/2$, and reasonable values of α $(\alpha < 0.25)$, $\dot{n}_{min}(d, \alpha) > n'_{min}(d, \alpha)$.

From (A3) follows that n'_{min} must be larger than 1. For practical purposes only values of d appreciably smaller than $0.5 - \alpha/2$ are of interest; hence n'_{min} will in practice always be considerably larger than 1.

A2. *Calculation of the Exact Confidence Coefficients in the HT-Procedure*

To prove that the HT-procedure actually provides the chosen confidence coefficients, one has to calculate for every test (d, α) the probability Prob $(\pi - d \leqslant \tilde{\pi} \leqslant \pi + d)$ of reaching decision points on or immediately outside the HT-curve, which provide estimates $\tilde{\pi}$, $\pi - d \leqslant \tilde{\pi} \leqslant \pi + d$. If this probability is $\geqslant (1 - 2\alpha)$, the observed interval estimate $[\tilde{\pi} - d, \tilde{\pi} + d]$ will cover the true parameter π with a confidence coefficient $\geqslant 1 - 2\alpha$.

Scheiber [5] has developed a method to calculate, for given π, the exact probabilities for arriving at a particular point (n, z) in the course of sequential sampling from a binomial distribution. For the HT-procedure, the probabilities of reaching a particular end point of a sample path on or immediately outside the decision curve need only be calculated for values of $z \geqslant 0$, because the probability of arriving at a point (n, z) when $\pi > 0.5$ is equal to the probability of reaching the symmetric point $(n, -z)$ when the true value of the parameter is $1 - \pi$.

For any decision curve (d, α) there exist a few points (n, z) immediately outside the curve, which may be arrived at from two points, $(n - 1, z - 1)$ and $(n - 1, z + 1)$. Theoretically this will affect the estimate $\tilde{\pi}$, defined as the intercept of the sample path with the decision curve. Due consideration has to be paid to this fact in calculating the exact confidence coefficients of the estimates $\tilde{\pi}$. In the following example there is only one indication for a possible influence of the sequential nature of sampling on the error probabilities, the upper tail of the error probability for $\pi = 0.65$ being slightly larger than α. This is caused by the first five decision points in Fig. 2 absorbing too many sample paths.

As an example for the HT-curve ($d = 0.15$, $\alpha = 0.025$) in Fig. 2 all the possible end points of sample paths are shown for $z \geqslant 0$, i.e. the decision points for the estimates $0.5 \leqslant \tilde{\pi} \leqslant 0.85$. The exact probabilities of arriving at the various decision points were calculated by the method of Scheiber [5] for $\pi = 0.15$ (0.05), 0.85. The results are summarized in Table 3.

Table 3
Exact probabilities of HT-estimates $\tilde{\pi}$.
The exact probabilities of arriving at correct $(\pi - d \leqslant \tilde{\pi} \leqslant \pi + d)$ and incorrect $(\tilde{\pi} < \pi - d, \tilde{\pi} > \pi + d)$ estimates $\tilde{\pi}$, based on the points on or immediately outside the HT-curve with $d = 0.15$, $\alpha = 0.025$ are given for different values of the true parameter π.

π	0.5	0.55	0.6	0.65	0.7	0.75	0.8	0.85
Prob $(\tilde{\pi} < \pi - d)$	.0165	.015	.017	.017	.017	.019	.016	.020
Prob $(\pi - d \leqslant \tilde{\pi} \leqslant \pi + d)$	.976	.969	.963	.956	.983	.981	.984	.980
Prob $(\tilde{\pi} > \pi + d)$	.0165	.016	.020	.027	–	–	–	–

As there are no decision points for $\tilde{\pi} > 1 - d$ $(\tilde{\pi} < d)$, the one-sided upper (lower) error probability is zero in these cases and the confidence coefficients become $\geqslant 1 - \alpha$. In the example given in Table 3 this is apparent for values of $\pi \geqslant 0.7$. Furthermore it can be seen that the confidence intervals obtained by the HT-procedure are not equally tailed for $\pi \neq 0.5$.

Calculations on Malý's curve ($d = 0.15$, $\alpha = 0.025$) lead to confidence coefficients of 0.995 ($\pi = 0.5$) and 0.997 ($\pi = 0.7$), both of which are much larger than the required value of 0.95.

A3. *Computer Simulation of Ties and Their Treatment*

Two independent normally distributed random variables X and Y with means μ_X, μ_Y and equal variance σ^2 were simulated, determining the distance between the two means so that Prob $(Y < X) = \pi$. Then the real axis was divided into intervals of length $4\,\sigma/a$, and any two 'measurements' of X and Y falling into the same interval were counted as a tie. Thus the 'accuracy' in the simulations is determined by a, for which 4 levels were chosen: $a = 10$, 100, 1000, and ∞. The highest accuracy, denoted by $a = \infty$, where no ties should occur, was in fact only that provided by the computer (8 significant digits); actually no ties were encountered in any simulation at that accuracy level.

2000 sample paths were simulated for various values of π, with ties treated according to the rule given in Section 6, and estimates $\tilde{\pi}$ based on the points on or immediately outside the HT-curve of Fig. 2 ($d = 0.15$, $\alpha = 0.025$). The relevant results obtained in these simulation experiments are:

1. The number of ties seems to be inversely proportional to the accuracy level a.
2. The observed confidence coefficients of the interval estimates $[\tilde{\pi} - d, \tilde{\pi} + d]$ are in good agreement with the exact probabilities (cf. Table 3), independent of the number of ties.

For two values of π (0.5, 0.85) the simulation results are summarized in Table 4, which permits a comparison with the exact probabilities in Table 3.

Table 4
Simulation results.
The simulation results of 2000 sample paths for $\pi = 0.5$ and $\pi = 0.85$ are summarized. The number of ties, correct ($\pi - d \leqslant \tilde{\pi} \leqslant \pi + d$) and incorrect ($\tilde{\pi} < \pi - d$, $\tilde{\pi} > \pi + d$) estimates $\tilde{\pi}$, based on the points on or immediately outside the HT-curve with $d = 0.15$, $\alpha = 0.025$, are given for 4 different levels of accuracy ($a = \infty$, 1000, 100, 10).

	$\pi = 0.5$				$\pi = 0.85$			
Accuracy (a)	∞	1000	100	10	∞	1000	100	10
Number of ties	0	105	999	10470	0	32	321	3214
$\tilde{\pi} < \pi - d$	34	33	31	25	42	43	50	37
$\pi - d \leqslant \tilde{\pi} \leqslant \pi + d$	1936	1937	1936	1953	1958	1957	1950	1963
$\tilde{\pi} > \pi + d$	30	30	33	22	–	–	–	–

References

[1] ARMITAGE, P. (1960): Sequential Medical Trials. Blackwell Scientific Publications, Oxford.

[2] BUKOVICS, E., und WOHLZOGEN, F. X. (1954): Biologische Auswertungen unter Verwendung von Sequentialtestverfahren. Zschr. Biol. *106*, 436–459.

[3] CLOPPER, C. J., and PEARSON, E. S. (1934): The Use of Confidence or Fiducial Limits Illustrated in the Case of the Binomial. Biometrika *26*, 404–413.

[4] MALY, V. (1960): Sequenzprobleme mit mehreren Entscheidungen und Sequenzschätzung. Biom. Zschr. *2*, 45–64.

[5] SCHEIBER, V. (1971): Berechnung der OC- und ASN-Funktionen geschlossener Sequentialtests bei Binomialverteilung. Computing *8*, 107–112.

[6] WALD, A. (1947): Sequential Analysis. John Wiley & Sons, New York.

[7] WOHLZOGEN, F. X. (1953): Quantitative Untersuchungen über Faktoren, die die Reaktion männlicher Kröten auf Choriongonadotropin beeinflussen. Arch. exper. Path. Pharmacol. *217*, 482–507.

[8] WOHLZOGEN, F. X., und WOHLZOGEN-BUKOVICS, E. (1966): Sequentielle Parameterschätzung bei biologischen Alles- oder Nichts-Reaktionen. Biom. Zschr. *8*, 83–120.

Author's address:
Institut für medizinische Statistik und Dokumentation, Universität Wien, Schwarzpanierstrasse 17, A–1090 Wien (Austria).

J. Pfanzagl

J. Pfanzagl

Investigating the Quantile of an Unknown Distribution

1. Introduction

If we want to estimate the q-quantile of an unknown distribution, can we do anything better than use the q-quantile of the sample? It does not seem unreasonable to expect an affirmative answer. For a distribution of known shape but unknown location (more formally: for a family of distributions with Lebesgue-density $x \to p(x - \theta), \theta \in \mathsf{R}$, p being known) we could use a maximum likelihood estimator which is asymptotically normal with an asymptotic variance smaller than that of the sample q-quantile. Since the sample also contains information about the shape of the distribution, it is not unreasonable to expect that the sample can be used to estimate first the shape of the distribution (taking here the part of a nuisance parameter), and to use this estimator to obtain an estimator for the q-quantile of the distribution which is superior to the sample q-quantile.

This general idea has been carried through successfully for the particular case of the median of a symmetric distribution: Even if the shape of the distribution is unknown, there exists a sequence of estimators with the same asymptotic variance as in the case in which the shape is known, and a corresponding result holds true for tests.

A rough sketch of an efficient test for the median of an unknown symmetric distribution has first been given by Stein ([17], Section 4). Starting from Hajeks [10] asymptotically efficient rank tests and an idea of Hodges and Lehmann [11] how to derive estimators from rank tests, van Eeden ([6], after a preliminary version in [5]) obtained an asymptotically efficient estimator for the median of an unknown symmetric distribution, based on ranks (Section 4 in [6] resp. Section 5 in [5]). Another estimator, also based on ranks, has been obtained independently by Weiss and Wolfowitz ([21], Section 7). This estimator is based on the trimmed sample, and its efficiency, depending on the trimming-point, can be brought arbitrarily close to one. Other papers containing related results are Takeuchi [19, 20], Fabian [7], Behran [2], Sacks [14] and Stone [18]. (For earlier results of this kind for the two-sample-problem see Bhattacharya [3].)

For the general problem of estimating an arbitrary quantile, an equally satisfactory solution cannot be expected. This becomes manifest if we consider a location and scale parameter family: It is only the case of a symmetric distribution in which the estimation of the (nuisance) scale parameter does not increase the asymptotic variance of the estimator of the location parameter. Nevertheless, it seems not unreasonable to expect that, in the general case of an arbitrary quantile of an arbitrary unknown distribution, there exists an estimator with an asymptotic variance somewhere between the asymptotic variance of the sample quantile and the asymptotic variance of the maximum likelihood estimator (based on the knowledge of the distribution). It is the purpose of this paper to show that this is not true: No translation-equivariant and asymptotically uniformly median unbiased estimator is asymptotically more concentrated about the distribution quantile than the sample-quantile. A corresponding result holds for tests.

All our optimality results refer to optimality up to terms of order $o(n^0)$, i.e. to efficiency. It is not unlikely that the quantiles cease to be optimal if terms of order $O(n^{-1/2})$ or $O(n^{-1})$ are considered.

It has been considered a triumph of nonparametric theory that—in the case of a symmetric distribution—the median can be estimated with the same asymptotic variance even if the shape of the distribution is unknown (in other words: that there exist nonparametric estimators of efficiency 1). Our results show that it is a somewhat restricted triumph, being limited to a case which is—considered properly—singular. This would not diminish the importance of this nonparametric result, if this singular case were of outstanding practical relevance. This, however, was never claimed. None of the papers cited above gives any motivation for the restriction to symmetric distributions, perhaps for good reasons: If we do not know anything about the distribution—how can we then know that it is symmetric?

Certainly there is a corresponding nonparametric result for the two-sample problem which holds without any restriction like symmetry: There exist tests and estimators of efficiency 1 for the difference in location of two identical (unknown) distributions (see Bhattacharya [3], van Eeden [5, 6], Weiss and Wolfowitz [21]). But if we do not know anything about the distributions occurring in the situation under investigation, how do we know that the 'treatment' effects the location only, and not the shape, too? Wouldn't it be more realistic to use a parametric model which includes enough nuisance parameters to allow for a great variety in shape as well as some amount of contamination, and gives us the chance to check whether the two distributions actually differ in location only?

2. The Results

Let $\mathscr{Q}$ denote the set of all p-measures (probability measures) over $\mathscr{B}$, the Borel-field of $\mathbf{R}$, which have an everywhere positive and differentiable Lebesque-density. For $Q \in \mathscr{Q}$, this (unique) density will be denoted by q. To obtain our results we have to assume that some properties (like 'level α' or 'median unbiasedness') hold uniformly on some set $\mathscr{P} \subset \mathscr{Q}$ of p-measures, and that the

p-measure P under investigation belongs to the interior of $\mathscr{P}$. In order to make the assertions applicable to as many elements of $\mathscr{P}$ as possible, the interior of $\mathscr{P}$ should be as large as possible. This suggests to endow $\mathscr{Q}$ with a large topology. A natural choice is the topology defined by the base

$$\left\{Q \in \mathscr{Q}: \sup_{x \in \mathbf{R}} \left| \frac{q(x)}{p(x)} - 1 \right| < \varepsilon \right\}, \quad \varepsilon > 0\,, \quad P \in \mathscr{Q}\,.$$

The interior of $\mathscr{P}$ with respect to this topology will be denoted by $\mathscr{P}^i$.

Restricted to the class of all p-measures $P \in \mathscr{Q}$ with bounded density, this topology is stronger than the topology of uniform convergence of densities. (Notice that p is bounded by the Lemma, whenever $\int (p'(x))^2/p(x)\, dx < \infty$.)

To illustrate the strength of this topology we mention that, for instance, $\mu_n \to \mu_0$ does not imply $N(\mu_n, \sigma^2) \to N(\mu_0, \sigma^2)$.

One possible application is to take $\mathscr{P} = \mathscr{Q}$. Another possible application is to contaminated distributions. Starting from some family $\mathscr{P}_0$ (e.g. the location parameter family of normal distributions with known variance), $\mathscr{P}$ is defined by $\mathscr{P}: = \{(1 - \alpha)\, P + \alpha\, Q: P \in \mathscr{P}_0,\, Q \in \mathscr{Q},\, \alpha \in [0, \alpha_0)\}$. (It is straightforward to verify that $\mathscr{P}$ is an open subset of $\mathscr{Q}$ if $\alpha_0 > 0$.)

For $q \in (0, 1)$, the q-quantile of a p-measure P will be denoted by $q(P)$. Given q, let

$$\mathscr{Q}_r: = \{Q \in \mathscr{Q}: q(Q) = r\}\,. \tag{1}$$

The set $\mathscr{Q}_r$ will always be endowed with the relative topology.

Furthermore, we shall need the following

Condition D: The p-measure $P \mid \mathscr{B}$ with Lebesgue-density p fulfills Condition D if there exists a function $M_P \mid \mathbf{R}$ with

$$\int M_P(x)^2\, p(x)\, dx < \infty \tag{2}$$

and a number $c_P > 0$ such that for all $x \in \mathbf{R}$ and all $|z - x| < c_P$, $|y - x| < c_P$

$$\left| \frac{p(z)}{p(y)} - 1 \right| \leq |z - y|\, M_P(x)\,. \tag{3}$$

Condition D is closely related to a condition for parametrized families of p-measures which was introduced by Daniels ([4], p. 152) and frequently used since (see e.g. Pfanzagl [13], p. 1501/1502).

Let $l: = \log p$. Then Condition D has the following immediate consequences.

If $l'(y)$ exists for $|y - x| < c_P$, then

$$|l'(y)| \leqq M_P(x), \tag{4}$$

since

$$l'(y) = \lim_{z \to y} \frac{p(z) - p(y)}{(z - y)\, p(y)}.$$

Furthermore, for all $x \in \mathbf{R}$ and all $|y - x| < c_P$,

$$\left| \frac{l(y) - l(x)}{y - x} \right| \leqq M_P(x) \tag{5}$$

which follows from (4) by the Mean Value Theorem.

Condition D implies, in particular, $\int \big(l'(x)\big)^2\, p(x)\, dx < \infty$, a condition needed for the maximum likelihood estimation of any location parameter, and it is perhaps not much stronger than this.

The following Theorem gives in formula (7) an upper bound for the power function against local alternatives, being valid for any sequence of critical functions which is uniformly asymptotic of level α for the hypothesis that the q-quantile equals r.

3. Theorem

Let $(\varphi_r^{(n)})_{n \in \mathbf{N}}$ be a sequence of critical functions which is uniformly asymptotic of level α for the hypothesis $\mathscr{P}_r \subset \mathscr{Q}_r$, i.e.

$$\overline{\lim_{n \to \infty}}\, \sup\{Q^n(\varphi_r^{(n)}) : Q \in \mathscr{P}_r\} \leqq \alpha. \tag{6}$$

Then for each $P \in \mathscr{P}_r^i$ (the interior of $\mathscr{P}_r$ in $\mathscr{Q}_r$) fulfilling Condition D,

$$\overline{\lim_{n \to \infty}}\, {}_{n^{-1/2}t}P^n(\varphi_r^{(n)}) \leqq \Phi(\Phi^{-1}(\alpha) + t\,\sigma), \tag{7}$$

where ${}_{n^{-1/2}t}P$ is the p-measure P shifted by the amount $n^{-1/2}\,t$ to the right, i.e.

${}_{n^{-1/2}t}P(B) = P\{x - n^{-1/2}\,t : x \in B\}$, $B \in \mathscr{B}$, and

$$\sigma := p(r)\, q^{-1/2}\, (1 - q)^{-1/2}. \tag{8}$$

The following Corollary 1 asserts that for any sequence of critical functions which is uniformly asymptotic of level α for the hypothesis that the q-quantile equals r, the power against local alternatives cannot exceed the power of the test based on an appropriate sample quantile.

Corollary 2 asserts that no sequence of translation invariant and uniformly asymptotically median unbiased estimators for the q-quantile is asymptotically more concentrated about the true value than the sample q-quantile.

Corollary 1: Let $\bar{q}_n := q + n^{-1/2}\,\Phi^{-1}(\alpha)\,q^{1/2}(1-q)^{1/2}$, *and let* $x^{(n)}_{\bar{q}_n}$ *denote the* $\bar{q}_n$*-quantile of the sample* $x_1, \ldots, x_n$. *Let*

$$C_r^{(n)} := \{(x_1, \ldots, x_n) \in \mathbf{R}^n : x^{(n)}_{\bar{q}_n} > r\}\,.$$

Under the assumptions of the Theorem:

(i) $P^n(C_r^{(n)})$, $n \in \mathbf{N}$, converges to α, uniformly for $P \in \mathcal{Q}_r$;

(ii) $\overline{\lim\limits_{n\to\infty}}\; {}_{n^{-1/2}t}P^n(\varphi_r^{(n)}) \leqq \lim\limits_{n\to\infty}\; {}_{n^{-1/2}t}P^n(C_r^{(n)})\,.$

Hence $C_r^{(n)}$ *is asymptotically optimal for testing the hypothesis* $q(P) = r$ *against the alternative* $q(P) > r$.

At first sight it might seem more natural to use the critical region

$$\{(x_1, \ldots, x_n) \in \mathbf{R}^n : x_q^{(n)} > r - n^{-1/2}\,\Phi^{-1}(\alpha)\,q^{1/2}(1-q)^{1/2}\,\hat{p}_n(r)^{-1}\},$$

where $\hat{p}_n(r)$ is an estimator of $p(r)$. But such a critical region will share the uniformity required by (6) only for particular choices of $\hat{p}_n(r)$.

Proof:

(i) Let E denote the uniform distribution over $(0, 1)$. If $q(P) = r$,

$$P^n(C_r^{(n)}) = P^n\{(x_1, \ldots, x_n) \in \mathbf{R}^n : x^{(n)}_{\bar{q}_n} > q(P)\}$$

$$= E^n\{(x_1, \ldots, x_n) \in (0, 1)^n : x^{(n)}_{\bar{q}_n} > q\}\,,$$

which converges to α by Lemma 2, Part I in Smirnov [16].

Since the last expression is independent of $P \in \mathcal{Q}_r$, the convergence is trivially uniform.

(ii) Let $F(t) := P(-\infty, t)$, $t \in \mathbf{R}$. For $P \in \mathcal{Q}_r$, we have $F(r) = q$ and therefore

$$F(r - n^{-1/2}t) = q - n^{-1/2}\,t\,p(r + \delta_n)$$

with $\delta_n \to 0$. Using this relation one easily obtains from Lemma 2, Part I in Smirnov [16] that

$$_{n^{-1/2}t}P^n(C_r^{(n)}) = P^n\{(x_1, \ldots, x_n) \in \mathbf{R}^n : x_{\bar{q}_n}^{(n)} > r - n^{-1/2}\, t\}$$

converges to $\Phi(\Phi^{-1}(\alpha) + t\, p(r)\, q^{-1/2}\, (1-q)^{-1/2})$. Together with (7) this implies the assertion.

Corollary 2: Let $(T^{(n)})_{n \in \mathbf{N}}$ *be a sequence of estimators for the q-quantile which is asymptotically median unbiased, uniformly over* $\mathscr{P} \subset \mathscr{Q}$, *i.e.*

$$\overline{\lim_{n\to\infty}} \sup_{P \in \mathscr{P}} P^n\{(x_1, \ldots, x_n) \in \mathbf{R}^n : T^{(n)}(x_1, \ldots, x_n) > q(P)\} \leqq \frac{1}{2}, \tag{9}$$

$$\overline{\lim_{n\to\infty}} \sup_{P \in \mathscr{P}} P^n\{(x_1, \ldots, x_n) \in \mathbf{R}^n : T^{(n)}(x_1, \ldots, x_n) < q(P)\} \leqq \frac{1}{2}. \tag{10}$$

Assume that for every $n \in \mathbf{N}$, $T^{(n)}$ *is equivariant under translations, i.e.*

$$T^{(n)}(x_1 + u, \ldots, x_n + u) = T^{(n)}(x_1, \ldots, x_n) + u\,,$$

$$x_i\,,\, u \in \mathbf{R}\,, \quad i = 1, \ldots, n\,.$$

Then for each $P \in \mathscr{P}^i$ *fulfilling Condition D and for arbitrary* $t', t'' \geqq 0$,

$$\overline{\lim_{n\to\infty}} P^n\{(x_1, \ldots, x_n) \in \mathbf{R}^n : q(P) - n^{-1/2}\, t' < T^{(n)}(x_1, \ldots, x_n) < q(P) + n^{-1/2}\, t''\}$$

$$\leqq \lim_{n\to\infty} P^n\{(x_1, \ldots, x_n) \in \mathbf{R}^n : q(P) - n^{-1/2}\, t' < x_q^{(n)} < q(P) + n^{-1/2}\, t''\}.$$

We remark that $x_q^{(n)}$ *is equivariant under translations and fulfills* (9) *and* (10) *uniformly for* $P \in \mathscr{Q}$.

Proof: Let $P \in \mathscr{P}^i$ be an arbitrary p-measure fulfilling Condition D. Let $\mathscr{P}_{q(P)} := \mathscr{P} \cap \mathscr{Q}_{q(P)}$ [see (1)]. Condition (9) implies

$$\overline{\lim_{n\to\infty}} \sup_{Q \in \mathscr{P}_{q(P)}} Q^n\{(x_1, \ldots, x_n) \in \mathbf{R}^n : T^{(n)}(x_1, \ldots, x_n) > q(P)\} \leqq \frac{1}{2}.$$

Let $(\mathscr{P}_{q(P)})^i$ denote the interior of $\mathscr{P}_{q(P)}$ in $\mathscr{Q}_{q(P)}$. Since

$$P \in \mathscr{P}^i \cap \mathscr{Q}_{q(P)} \subset (\mathscr{P}_{q(P)})^i$$

(see Gaal [8], p. 55, Lemma 1 (iv)), the Theorem is applicable for

$$\varphi_{q(P)}^{(n)} = 1_{\{T^{(n)} > q(P)\}} \quad \text{and} \quad \alpha = \frac{1}{2}.$$

We obtain from (7)

$$\overline{\lim_{n\to\infty}}\ {}_{n^{-1/2}t'}P^n\{(x_1, \ldots, x_n) \in \mathsf{R}^n : T^{(n)}(x_1, \ldots, x_n) > q(P)\} \leqq \Phi(t'\sigma).$$

Since $T^{(n)}$ is equivariant, we have

$$\begin{aligned}
&{}_{n^{-1/2}t'}P^n\{(x_1, \ldots, x_n) \in \mathsf{R}^n : T^{(n)}(x_1, \ldots, x_n) > q(P)\}\\
&\quad = P^n\{(x_1, \ldots, x_n) \in \mathsf{R}^n : T^{(n)}(x_1 + n^{-1/2}t', \ldots, x_n + n^{-1/2}t') > q(P)\}\\
&\quad = P^n\{(x_1, \ldots, x_n) \in \mathsf{R}^n : T^{(n)}(x_1, \ldots, x_n) > q(P) - n^{-1/2}t'\},
\end{aligned}$$

hence

$$\overline{\lim_{n\to\infty}} P^n\{(x_1, \ldots, x_n) \in \mathsf{R}^n : T^{(n)}(x_1, \ldots, x_n) > q(P) - n^{-1/2}t'\} \leqq \Phi(t'\sigma)$$

and therefore

$$\begin{aligned}
&\underline{\lim_{n\to\infty}} P^n\{(x_1, \ldots, x_n) \in \mathsf{R}^n : T^{(n)}(x_1, \ldots, x_n) \leqq q(P) - n^{-1/2}t'\}\\
&\quad \geqq 1 - \Phi(t'\sigma) = \Phi(-t'\sigma). \qquad (11)
\end{aligned}$$

Starting from (10) we obtain by the corresponding argument

$$\begin{aligned}
&\underline{\lim_{n\to\infty}} P^n\{(x_1, \ldots, x_n) \in \mathsf{R}^n : T^{(n)}(x_1, \ldots, x_n) \geqq q(P) + n^{-1/2}t''\}\\
&\quad \geqq \Phi(-t''\sigma). \qquad (12)
\end{aligned}$$

(11) and (12) together imply

$$\begin{aligned}
&\overline{\lim_{n\to\infty}} P^n\{(x_1, \ldots, x_n) \in \mathsf{R}^n : q(P) - n^{-1/2}t' < T^{(n)}(x_1, \ldots, x_n)\\
&\qquad\qquad < q(P) + n^{-1/2}t''\}\\
&\quad \leqq \int_{-t'\sigma}^{t''\sigma} \varphi(u)\, du\\
&\quad = \lim_{n\to\infty} P^n\{(x_1, \ldots, x_n) \in \mathsf{R}^n : q(P) - n^{-1/2}t' < x_q^{(n)} < q(P) + n^{-1/2}t''\}.
\end{aligned}$$

Proof of the Theorem:

(i) Let a p-measure $P \in \mathscr{P}_r^i$ fulfilling Condition D be given. Let p denote a differentiable density of P, and let $l := \log p$.

At first, we remark that

$$\int_{-\infty}^{z} l'(x)\, p(x)\, dx = p(z) \quad \text{and} \quad \int_{z}^{\infty} l'(x)\, p(x)\, dx = -p(z)\,, \quad z \in \mathbf{R}\,. \tag{13}$$

This can be seen as follows: Let $\varepsilon > 0$ be arbitrary. Since l' is P-integrable by (4) and (2), there exists z_ε such that

$$\int_{-\infty}^{z_\varepsilon} | l'(x) |\, p(x)\, dx < \varepsilon/2\,.$$

Since $p(x)$ goes to zero as x goes to $-\infty$ (cf. the Lemma), z_ε can be chosen such that $p(z_\varepsilon) < \varepsilon/2$.

Hence,

$$\left| \int_{-\infty}^{z} l'(x)\, p(x)\, dx - p(z) \right| = \left| \int_{-\infty}^{z_\varepsilon} l'(x)\, p(x)\, dx + \int_{z_\varepsilon}^{z} p'(x)\, dx - p(z) \right|$$

$$= \left| \int_{-\infty}^{z_\varepsilon} l'(x)\, p(x)\, dx - p(z_\varepsilon) \right| < \varepsilon\,.$$

The last equality follows (see Natanson [12], VIII, § 2, Theorem 5) from the fact that p' exists everywhere and is integrable with respect to the Lebesgue measure by (4) and (2).

Let

$$\Delta_{n,t}(x) := n^{1/2}\, t^{-1}\big(l(x) - l(x - n^{-1/2}\, t)\big)\,. \tag{14}$$

By (5) we have for all $x \in \mathbf{R}$, $n \geqq n_t$

$$\max\{| \Delta_{n,t}(x) |\,,\ | \Delta_{n,t}(x + n^{-1/2}\, t) |\} \leqq M_P(x)\,. \tag{15}$$

(Here and in the following, n_t will be used as a generic symbol, i.e. the n_t's occurring in the different places are not necessarily identical.)

Furthermore,

$$\lim_{n\to\infty} \Delta_{n,t}(x) = \lim_{n\to\infty} \Delta_{n,t}(x + n^{-1/2}\, t) = l'(x)\,, \quad x \in \mathbf{R}\,. \tag{16}$$

Let $b_n \mid \mathbf{R}$ be a sequence of nondecreasing differentiable functions with the following properties:

$$| b_n(u) | \leqq \min\{| u |, n^{1/4}\}, \tag{17}$$

$$\lim_{n\to\infty} b_n(u) = u . \tag{18}$$

By (15) and (17)

$$\max\{| b_n(\Delta_{n,t}(x)) |, \; | b_n(\Delta_{n,t}(x + n^{-1/2} t)) |\} \leqq M_P(x) . \tag{19}$$

Since the functions b_n are monotone and the limit function is continuous, the convergence in (18) is continuous. (For a proof see Lemma 1.2 in Schmetterer [15].) Hence (16) implies

$$\lim_{n\to\infty} b_n(\Delta_{n,t}(x)) = \lim_{n\to\infty} b_n(\Delta_{n,t}(x + n^{-1/2} t)) = l'(x), \quad x \in \mathbf{R} . \tag{20}$$

By the Bounded Convergence Theorem (which is applicable by (19) and (2)), we obtain from (20) and (13),

$$\begin{aligned} \beta_n &:= \int_{-\infty}^{r} b_n(\Delta_{n,t}(x))\, p(x)\, dx = p(r) + o(n^0), \\ \beta_n' &:= \int_{r}^{\infty} b_n(\Delta_{n,t}(x))\, p(x)\, dx = -\, p(r) + o(n^0) . \end{aligned} \tag{21}$$

Let

$$k_0(x) := \begin{cases} p(r)\, q^{-1} & x \leqq r \\ -\, p(r)\, (1-q)^{-1} & x > r . \end{cases} \tag{22}$$

From (22) and (8),

$$\int k_0(x)\, p(x)\, dx = 0 , \tag{23}$$

$$\int k_0(x)^2\, p(x)\, dx = \sigma^2 . \tag{24}$$

Since $P \in \mathscr{P}_r$ we have

$$\lim_{n\to\infty} \beta_n = \int_{-\infty}^{r} k_0(x)\, p(x)\, dx \quad \text{and} \quad \lim_{n\to\infty} \beta'_n = \int_{r}^{\infty} k_0(x)\, p(x)\, dx .$$

Hence there exists a sequence of nonincreasing differentiable functions $k_n \mid \mathbf{R}$, $n \in \mathbf{N}$, with the following properties:

$$c := \sup\{|k_n(x)| : x \in \mathbf{R}, n \in \mathbf{N}\} < \infty , \tag{25}$$

$$\int_{-\infty}^{r} k_n(x)\, p(x)\, dx = \beta_n , \quad \int_{r}^{\infty} k_n(x)\, p(x)\, dx = \beta'_n , \quad n \in \mathbf{N} , \tag{26}$$

$$\lim_{n\to\infty} k_n(x) = k_0(x) , \quad x \in \mathbf{R} . \tag{27}$$

Now we define

$$g_n(x) := k_n(x) - b_n\big(\Delta_{n,t}(x)\big) , \quad x \in \mathbf{R} , \quad n \in \mathbf{N} . \tag{28}$$

From (15), (17) and (25) we have for all $x \in \mathbf{R}$, $n \geqq n_t$,

$$\max\{|g_n(x)|, |g_n(x + n^{-1/2} t)|\} \leqq \min\{M_P(x), n^{1/4}\} + c . \tag{29}$$

From (21) and (26) we have for all $n \in \mathbf{N}$,

$$\int_{-\infty}^{r} g_n(x)\, p(x)\, dx = 0 , \quad \int_{r}^{\infty} g_n(x)\, p(x)\, dx = 0 . \tag{30}$$

Since k_n is a monotone function, the sequence k_n, $n \in \mathbf{N}$, converges to k_0 continuously with the possible exception of r, where k_0 is discontinuous. (For a proof see Lemma 1.2 in Schmetterer [15].)

Together with (27) and (20), relation (28) therefore implies

$$\lim_{n\to\infty} g_n(x) = \lim_{n\to\infty} g_n(x + n^{-1/2} t) = k_0(x) - l'(x) , \quad x \in \mathbf{R} . \tag{31}$$

Finally, let

$$p_n(x) := p(x)\big(1 + n^{-1/2} t\, g_n(x)\big) , \quad n \in \mathbf{N} , \quad x \in \mathbf{R} . \tag{32}$$

Because of (29), p_n is positive for $n \geqq n_t$ and

$$p_n(x) \leqq 2\, p(x)\,, \quad x \in \mathbf{R}\,, \quad n \geqq n_t\,. \tag{33}$$

By (30) p_n, $n \geqq n_t$, is the density of a p-measure, say P_n, with q-quantile r. (We remark that the definition of p_n given in (32) is suggested by the Euler-equation of the isoperimetric problem. See, e.g., Gelfand and Fomin [9], p. 43, Theorem 1.) Furthermore, g_n is differentiable, and so is p_n. Hence for all $n \geqq n_t$, $P_n \in \mathcal{Q}_r$ [see (1)]. Furthermore, (32) and (29) imply

$$\lim_{n\to\infty} \sup_{x\in\mathbf{R}} \left| \frac{p_n(x)}{p(x)} - 1 \right| = 0\,. \tag{34}$$

Since $P \in \mathscr{P}_r^i$, this implies

$$P_n \in \mathscr{P}_r \quad \text{for} \quad n \geqq n_t\,. \tag{35}$$

(ii) For notational convenience let

$$f_n(x) := n^{1/2}\, t^{-1}\big(l(x - n^{-1/2}\, t) - \log p_n(x)\big)\,, \quad x \in \mathbf{R}\,, \quad n \in \mathbf{N}\,. \tag{36}$$

We define

$$\mu_n := \int f_n(x)\, p_n(x)\, dx\,, \tag{37}$$

$$\sigma_n^2 := \int \big(f_n(x) - \mu_n\big)^2\, p_n(x)\, dx\,, \tag{38}$$

$$\bar{\mu}_n := \int f_n(x)\, p(x - n^{-1/2}\, t)\, dx\,, \tag{39}$$

$$\bar{\sigma}_n^2 := \int \big(f_n(x) - \bar{\mu}_n\big)^2\, p(x - n^{-1/2}\, t)\, dx\,. \tag{40}$$

We have

$$f_n(x) = -\Delta_{n,t}(x) - n^{1/2}\, t^{-1} \log\big(1 + n^{-1/2}\, t\, g_n(x)\big)\,. \tag{41}$$

Since $|u| \leqq 1/2$ implies $|\log(1+u)| \leqq 2\,|u|$, we obtain from (41), (15), and (29) for $n \geqq n_t$

$$\max\{|f_n(x)|\,, \ |f_n(x + n^{-1/2}\, t)|\} \leqq 3\, M_P(x) + 2\, c\,. \tag{42}$$

Using

$$\lim_{n\to\infty} n^{1/2}\, t^{-1} \log\bigl(1 + n^{-1/2}\, t\, g_n(x)\bigr) = \lim_{n\to\infty} g_n(x)$$

we obtain from (41), (16), and (31),

$$\lim_{n\to\infty} f_n(x) = \lim_{n\to\infty} f_n(x + n^{-1/2}\, t) = -\, k_0(x)\,, \quad x \in \mathbf{R}\,. \tag{43}$$

Using the Bounded Convergence Theorem [which is applicable by (42), (2), and (33)] we obtain from (37) to (40), (43), (34), (24), and (23),

$$\lim_{n\to\infty} \mu_n = \lim_{n\to\infty} \bar{\mu}_n = 0$$

and

$$\lim_{n\to\infty} \sigma_n^2 = \lim_{n\to\infty} \bar{\sigma}_n^2 = \sigma^2\,. \tag{44}$$

$$\Bigl[\text{Hint: Use } \bar{\mu}_n = \int f_n(x + n^{-1/2}\, t)\, p\,(x)\, dx \quad \text{and}$$

$$\bar{\sigma}_n^2 = \int \bigl(f_n(x + n^{-1/2}\, t) - \bar{\mu}_n\bigr)^2\, p(x)\, dx\Bigr]\,.$$

We have

$$n^{1/2}\, t^{-1}\bigl(p_n(x) - p(x - n^{-1/2}\, t)\bigr)$$

$$= n^{1/2}\, t^{-1}\bigl(p(x) - p(x - n^{-1/2}\, t)\bigr) + p(x)\, g_n(x)\,. \tag{45}$$

By (45), (3), and (29)

$$\bigl|\, n^{1/2}\, t^{-1}\bigl(p_n(x) - p(x - n^{-1/2}\, t)\bigr)\bigr| \leqq \bigl(2\, M_P(x) + c\bigr)\, p(x)\,. \tag{46}$$

By (45) and (31)

$$\lim_{n\to\infty} n^{1/2}\, t^{-1}\bigl(p_n(x) - p(x - n^{-1/2}\, t)\bigr) = k_0(x)\, p(x)\,. \tag{47}$$

Using (37), (39), (43), (47), (24) and the Bounded Convergence Theorem [which applies because of (42), (46) and (2)] we obtain

$$\lim_{n\to\infty} n^{1/2}(\mu_n - \bar{\mu}_n) = -\, t\, \sigma^2\,. \tag{48}$$

By arguments of the same kind we obtain that for each $\varepsilon > 0$

$$\int\limits_{\{y \in \mathbf{R}: (f_n(y) - \mu_n)^2 > \varepsilon n c_n^2\}} (f_n(x) - \mu_n)^2 \, p_n(x) \, dx = o(n^0) \tag{49}$$

and

$$\int\limits_{\{y \in \mathbf{R}: (f_n(y) - \bar{\mu}_n)^2 > \varepsilon n \bar{\sigma}_n^2\}} (f_n(x) - \bar{\mu}_n)^2 \, p(x - n^{-1/2} t) \, dx = o(n^0) \, . \tag{50}$$

(iii) After these preliminaries, the proof can be concluded as follows[1]:

For $n \in \mathbf{N}$, $u \in \mathbf{R}$ let

$$C_n(u) := \{(x_1, \ldots, x_n) \in \mathbf{R}^n : n^{-1/2} \sigma_n^{-1} \sum_{i=1}^{n} \big(f_n(x_i) - \mu_n\big) \geqq u\} \, . \tag{51}$$

(49) implies that the sequence of p-measures induced on $\mathscr{B}^n$ by $P_n^n \mid \mathscr{B}^n$ and the map $(x_1, \ldots, x_n) \to \big(f_n(x_1), \ldots, f_n(x_n)\big)$ fulfills Lindeberg's condition.

Hence

$$\lim_{n \to \infty} P_n^n\big(C_n(u)\big) = 1 - \Phi(u) \, , \quad u \in \mathbf{R} \, . \tag{52}$$

(Notice that Lindeberg's condition is also sufficient in the case of triangular arrays (see Billingsley [1], Theorem 7.2, p. 42).)

Using (6) and (35) we obtain

$$\overline{\lim_{n \to \infty}} \, P_n^n(\varphi_r^{(n)}) \leqq \alpha \, . \tag{53}$$

1) Some readers will be tormented by the question why we do not use the theory of contiguity. The reason is that we would then have to prove that

$$\mu_n = -\frac{1}{2} n^{-1/2} t \, \sigma^2 + o(n^{-1/2})$$

(which is, in fact, true) and that the sequences P^n and P_n^n, $n \in \mathbf{N}$, are contiguous (which is doubtful). In any case, the proof would not be simplified by using contiguity.

Let $v > \Phi^{-1}(\alpha)$ be arbitrary. By (52),

$$\lim_{n\to\infty} P_n^n(C_n(-v)) = \Phi(v) > \alpha .$$

Together with (53) this implies for all $n \geqq n_t$

$$P_n^n(\varphi_r^{(n)}) \leqq P_n^n(C_n(-v)) . \tag{54}$$

By the Fundamental Lemma of Neyman-Pearson, $C_n(-v)$ is most powerful for testing the hypothesis P_n^n against the alternative ${}_{n^{-1/2}t}P^n$ [see (36) and (51)]. Hence (54) implies for all $n \geqq n_t$

$$\begin{aligned} {}_{n^{-1/2}t}P^n(\varphi_r^{(n)}) &\leqq {}_{n^{-1/2}t}P^n(C_n(-v)) \\ &= {}_{n^{-1/2}t}P^n\{(x_1, \ldots, x_n) \in \mathbf{R}^n : n^{-1/2}\,\bar{\sigma}_n^{-1} \sum_{i=1}^n (f_n(x_i) - \bar{\mu}_n) \\ &\qquad \geqq -\bar{\sigma}_n^{-1}\,\sigma_n\,v + \bar{\sigma}_n^{-1}\,n^{1/2}(\mu_n - \bar{\mu}_n)\} . \end{aligned} \tag{55}$$

From (44) and (48) we obtain

$$\lim_{n\to\infty}[-\bar{\sigma}_n^{-1}\,\sigma_n\,v + \bar{\sigma}_n^{-1}\,n^{1/2}(\mu_n - \bar{\mu}_n)] = -v - t\,\sigma . \tag{56}$$

By (50), Lindeberg's condition is fulfilled for the sequence of p-measures induced on $\mathscr{B}^n$ by ${}_{n^{-1/2}t}P^n \mid \mathscr{B}^n$ and the map $(x_1, \ldots, x_n) \to (f_n(x_1), \ldots, f_n(x_n))$. Hence

$$u \to {}_{n^{-1/2}t}P^n\{(x_1, \ldots, x_n) \in \mathbf{R}^n : n^{-1/2}\,\bar{\sigma}_n^{-1} \sum_{i=1}^n (f_n(x_i) - \bar{\mu}_n) \geqq u\}$$

converges continuously to $1 - \Phi$.

Together with (55) and (46) this implies

$$\overline{\lim_{n\to\infty}}\; {}_{n^{-1/2}t}P^n(\varphi_r^{(n)}) \leqq 1 - \Phi(-v - t\,\sigma) = \Phi(v + t\,\sigma) . \tag{57}$$

Since $v > \Phi^{-1}(\alpha)$ was arbitrary, (57) implies (7).

The following Lemma entails that the asymptotic variance of a sample quantile is never smaller than the asymptotic variance of the maximum likeli hood estimator of a location parameter. It also implies that the density vanishes in the neighborhood of $\pm\infty$.

Lemma: Let p be the Lebesgue-density of a p-measure; assume that p' exists everywhere. Then

$$\int \frac{(p'(x))^2}{p(x)}\, dx \geqq \frac{p(r)^2}{\int\limits_{-\infty}^{r} p(x)\, dx \int\limits_{r}^{\infty} p(x)\, dx} \quad \text{for all} \quad r \in \mathbf{R}\,.$$

Proof: Given $r \in \mathbf{R}$, let $g(x) := \begin{cases} \int\limits_{r}^{\infty} p(y)\, dy & x \leqq r \\ -\int\limits_{-\infty}^{r} p(y)\, dy & x > r \end{cases}$.

The assertion follows easily from

$$\left(\int\limits_{-\infty}^{+\infty} g(x)\, p'(x)\, dx\right)^2 \leqq \int\limits_{-\infty}^{+\infty} g(x)^2\, p(x)\, dx \int\limits_{-\infty}^{+\infty} \frac{(p'(x))^2}{p(x)}\, dx\,.$$

4. Acknowledgment

The author wishes to thank Mr. R. Michel for the careful examination of the manuscript and for suggestions which led to substantial simplifications in the proof of the Theorem. Furthermore, he wishes to thank Mr. R. Reiss for a critical remark.

References

[1] BILLINGSLEY, P. (1968): Convergence of Probability Measures. Wiley, New York.

[2] BEHRAN, R. (1974): Asymptotically efficient adaptive rank estimates in location models. Ann. Statist. *2*, 63–74.

[3] BHATTACHARYA, P. K. (1967): Efficient estimation of a shift parameter from grouped data. Ann. Math. Statist. *38*, 1770–1787.

[4] DANIELS, H. E. (1961): The asymptotic efficiency of a maximum likelihood estimator. Proc. Fourth Berkeley Symp. on Math. Statist. and Prob. *1*, 151–163.

[5] VAN EEDEN, C. (1968): Nonparametric Estimation. Seventh Session of the 'Seminaire Mathématiques Supérieures'. Les Presses de l'Université de Montréal.

[6] VAN EEDEN, C. (1970): Efficiency–robust estimation of location. Ann. Math. Statist. *41*, 172–181.

[7] FABIAN, V. (1973): Asymptotically efficient stochastic approximation: the RM case. Ann. Statist. *1*, 486–495.

[8] GAAL, S. A. (1964): Point Set Topology. Academic Press, New York and London.

[9] GELFAND, I. M. and FOMIN, S. V. (1963): Calculus of Variations. Prentice-Hall, Englewood Cliffs.

[10] HAJEK, J. (1962): Asymptotically most powerful rank order tests. Ann. Math. Statist. *33*, 1124–1147.

[11] HODGES, J. L. and LEHMANN, E. (1963): Estimates of location based on rank tests. Ann. Math. Statist. *34*, 598–611.

[12] NATANSON, I. P. (1954): Theorie der Funktionen einer reellen Veränderlichen. Akademie-Verlag, Berlin.

[13] PFANZAGL, J. (1970): On the asymptotic efficiency of median unbiased estimates. Ann. Math. Statist. *41*, 1500–1509.

[14] SACKS, J. (1974): An asymptotically efficient sequence of estimators of a location parameter. Unpublished manuscript.

[15] SCHMETTERER, L. (1966): On the asymptotic efficiency of estimates. Research Papers in Statistics. Festschrift for J. Neyman (ed. by F. N. David). Wiley, London.

[16] SMIRNOV, N. V. (1949): The limit distributions for the terms of a variational series. Trudy Mat. Inst. Steklov *25* (in Russian). English translation in Amer. Math. Soc. Transl. (1) *11* (1962), 82–143.

[17] STEIN, C. (1956): Efficient nonparametric testing and estimation. Proc. Third Berkeley Symp. on Math. Statist. and Prob. *1*, 187–195.

[18] STONE, C. J. (1974): Empirical approximate maximum likelihood estimators of a location parameter. Unpublished manuscript.

[19] TAKEUCHI, K. (1969): A uniformly asymptotically efficient robust estimator of a location parameter. Report IMM 375, Courant Institute of Mathematical Sciences, New York University.

[20] TAKEUCHI, K. (1971): A uniformly asymptotically efficient estimator of a location parameter. J. Amer. Statist. Ass. *66*, 292–301.

[21] WEISS, L. and WOLFOWITZ, J. (1970): Asymptotically efficient non-parametric estimators of location and scale parameters. Z. Wahrscheinlichkeitstheorie verw. Geb. *16*, 134–150.

Author's address:
J. Pfanzagl, University of Cologne.

4. Testing

M. L. Puri
and C. Radhakrishna Rao

C. Radhakrishna Rao

Augmenting Shapiro-Wilk Test for Normality[1])

Summary

The expected value of the i-th order statistic is considered to be $\gamma_0 + \gamma_1 c_i + \gamma_2(c_i^2 - \lambda) + \gamma_3(c_i^3 - \mu c_i) + \cdots$, where c_i is the expected value of the i-th order statistic from $N(0, 1)$. If the underlying distribution is normal, then $\gamma_1 = \sigma$, $\gamma_2 = 0$, $\gamma_3 = 0, \ldots$. Shapiro-Wilk test is designed to test the hypothesis $\gamma_1 = \sigma$. The possibility of testing the multiple hypothesis $\gamma_1 = \sigma$, $\gamma_2 = 0$, $\gamma_3 = 0, \ldots$, is considered in the present paper. Some improvement in efficiency over the Shapiro-Wilk test is demonstrated.

1. Introduction

A number of investigations have been made during recent years to compare the power properties of various tests for normality, an excellent summary of which is given by Stephens [11] along with his own contributions to the subject. All these studies indicate that the W test introduced by Shapiro and Wilk [9] and further examined by Shapiro, Wilk and Chen [8] provides an omnibus test for normality, which is generally more powerful than other known tests for a wide variety of departures from normality.

The basis of the W test is as follows. Let $\boldsymbol{Y}' = (Y_1, \ldots, Y_n)$ be the vector of order statistics in a sample of size n from an unknown population. Further let $\boldsymbol{c}' = (c_1, \ldots, c_n) = E(\boldsymbol{Y}')$ when the population is $N(0, 1)$. It is well known that when the population is $N(\mu, \sigma^2)$

$$E(\boldsymbol{Y}) = \mu\, \boldsymbol{c}_0 + \sigma\, \boldsymbol{c}\,, \tag{1.1}$$

1) Research supported by the National Science Foundation Grant GP 27715 and Air Force Office of Scientific Research AFSC, USAF, under grant number AFOSR 71–2009.

where $\boldsymbol{c}_0' = (1, \ldots, 1)$. If the dispersion matrix of $\boldsymbol{Y}$ is $\sigma^2 \boldsymbol{V}$, then the Blue of σ, using the Gauss-Markoff theory (see Rao [6]) is

$$\sigma^* = \boldsymbol{c}' \boldsymbol{V}^{-1} \boldsymbol{Y} \div \boldsymbol{c}' \boldsymbol{V}^{-1} \boldsymbol{c} \, . \tag{1.2}$$

But an unbiased estimator of σ^2 under any hypothesis is $(\widehat{\sigma})^2 = \Sigma(Y_i - \bar{Y})^2 \div (n-1)$, provided the second moment of the population exists. Then a test of the model is provided by the closeness of the ratio $(\sigma^*/\widehat{\sigma})^2$ to unity. The statistic W, as defined by Shapiro and Wilk, is the ratio $(\sigma^*/\widehat{\sigma})^2$ apart from a constant multiplier, and the test criterion is $W \leq W_p$, where W_p is the lower p-percent point of the distribution of W on the normal hypothesis.

In the present paper we explore the possibility of comparing the model (1.1) with an alternative model of the form

$$E(Y_i) = \gamma_1 + \gamma_2 \, c_i + \gamma_3(c_i^2 - \lambda) + \gamma_4(c_i^3 - \mu \, c_i) + \cdots, \tag{1.2}$$

where λ and μ are constants chosen to provide orthogonal polynomials. More specifically, we fit the model (1.2) to observed data and compare the estimated coefficients $\gamma_2^*, \gamma_3^*, \gamma_4^*, \ldots$, standardized by $\widehat{\sigma}$, with the theoretical values $1, 0, 0, \ldots$. The W test corresponds to the comparison of $\gamma_2^*/\widehat{\sigma}$ with 1. We examine whether any additional information is gained by augumenting the W test with tests on γ_3^* and γ_4^*. The possibility of such a procedure was examined by Heckard [2] at the suggestion made by one of the authors (Rao) at a lecture delivered at the Pennsylvania State University in 1970, but the study was incomplete.

Shapiro and Wilk [9] also mention the possibility of fitting a higher degree polynomial as in (1.2) but did not indicate how exactly the estimated coefficients can be used in tests of significance.

In our approach we accept W as the basic overall test and examine whether further information is gained by augmenting it with departures in γ_3^* and γ_4^* from zero.

The W test when significant does not by itself indicate the nature of departure from normality. We may then calculate γ_3^*. If its absolute value is large, then asymmetry is indicated. Similarly if γ_4^* is large, deviation from normal kurtosis is indicated. Thus supplementing the W test by tests based on γ_3^* and γ_4^* is useful in exploratory research.

On the other hand we may wish to devise a test of normality when the possible alternatives are skew distributions. In such a case it appears that W by itself or a test using γ_3^* alone is less powerful than the tests $T_{23(u)}$, $T_{23(l)}$ and T_{23} (defined in Table 1) based on combinations of W and γ_3^*.

Similarly, as tests of normality when the possible alternatives differ from normal kurtosis, the statistics $T_{24(u)}$, $T_{24(l)}$ and T_{24} (defined in Table 1) based on a combination of W and γ_4^* are more powerful than W or γ_4 alone.

It may be noted that our study departs in some sense from the type discussed by Durbin and Knott [1] and originated by Neyman [5].

A test for normality comparable to that of Shapiro-Wilk is given by Linder (see Rao, Matthai and Mitra [7]) based on the technique of fractile graphical analysis (FGA) developed by Mahalanobis [4]. Monte Carlo studies on Linder's test are reported in Linder and Czegledy [3]. Our approach can be adopted to augument Linder's test by fitting a polynomial to the fractile graph.

For references to previous literature on the subject, the reader is referred to papers by Shapiro and Wilk [9], Shapiro, Wilk and Chen [8], Durbin and Knott [1], Linder and Czegledy [3] and Stephens [11].

2. Derivation of Test Criteria

As explained in the introduction, we consider the model (1.2) as an alternative to the model (1.1). We choose only the first four terms for illustrating the test procedures. Defining the vetctors

$$\boldsymbol{c}_j' = (c_1^j, \ldots, c_n^j), \quad j = 0, 1, 2, \ldots,$$

$$\boldsymbol{b}_1 = \boldsymbol{c}_0, \quad \boldsymbol{b}_2 = \boldsymbol{c}_1,$$

$$\boldsymbol{b}_3 = \boldsymbol{c}_2 - \frac{\boldsymbol{c}_2' \boldsymbol{V}^{-1} \boldsymbol{c}_0}{\boldsymbol{c}_0' \boldsymbol{V}^{-1} \boldsymbol{c}_0}, \quad \boldsymbol{b}_4 = \boldsymbol{c}_3 - \frac{\boldsymbol{c}_3' \boldsymbol{V}^{-1} \boldsymbol{c}_1}{\boldsymbol{c}_1' \boldsymbol{V}^{-1} \boldsymbol{c}_1},$$

where $\boldsymbol{V}$ is the dispersion matrix of order statistics under $N(0, 1)$, we write (1.2), considering only four terms as

$$E(\boldsymbol{Y}) = \gamma_1 \boldsymbol{b}_1 + \gamma_2 \boldsymbol{b}_2 + \gamma_3 \boldsymbol{b}_3 + \gamma_4 \boldsymbol{b}_4 . \tag{2.1}$$

The *Blue*'s of the γ coefficients are

$$\gamma_i^* = \boldsymbol{b}_i' \boldsymbol{V}^{-1} \boldsymbol{Y} \div \boldsymbol{b}_i' \boldsymbol{V}^{-1} \boldsymbol{b}_i', \tag{2.2}$$

$$V(\gamma_i^*) = \frac{\sigma^2}{\boldsymbol{b}_i' \boldsymbol{V}^{-1} \boldsymbol{b}_i}. \tag{2.3}$$

Denoting $S^2 = \Sigma(Y_i - \bar{Y})^2$ and $\hat{\sigma}^2 = S^2/(n-1)$, we define the statistics

$$T_2 = W^{1/2} = \gamma_2^*(\boldsymbol{b}_2' \boldsymbol{V}^{-1} \boldsymbol{b}_2) \div \hat{\sigma}(\boldsymbol{b}_2' \boldsymbol{V}^{-1} \boldsymbol{V}^{-1} \boldsymbol{b}_2)^{1/2}, \tag{2.4}$$

$$T_3 = \gamma_3^*(\boldsymbol{b}_3' \boldsymbol{V}^{-1} \boldsymbol{b}_3)^{1/2} \div \hat{\sigma}, \tag{2.5}$$

$$T_4 = \gamma_4^*(\boldsymbol{b}_4' \boldsymbol{V}^{-1} \boldsymbol{b}_4)^{1/2} \div \hat{\sigma}. \tag{2.6}$$

The test $W^{1/2}$ (represented by T_2) is the square root of the Shapiro-Wilk W.

It may be seen that departures in T_3 and T_4 from the theoretical value zero indicate asymmetry and non-normal kurtosis respectively.

With the three statistics T_2, T_3 and T_4, one is faced with the problem of using them jointly in a suitable test procedure. We propose the following which accepts T_2 as the basic test and utilizes the additional information given by T_3 and T_4.

Table 1
Test Procedures.

Statistic	Test
$T_2 = W^{1/2}$	$T_2 \leq p_2$
$T_{23(u)}$	$T_2 \leq p_2$ or $T_3 \geq p_3$
$T_{23(l)}$	$T_2 \leq p_2$ or $T_3 \leq -p_3$
T_{23}	$T_2 \leq p_2$ or $\lvert T_3 \rvert \geq p_3'$
$T_{24(u)}$	$T_2 \leq p_2$ or $T_4 \geq p_4$
$T_{24(l)}$	$T_2 \leq p_2$ or $T_4 \leq -p_4$
T_{24}	$T_2 \leq p_2$ or $\lvert T_4 \rvert \geq p_4'$

In the above the suffixes u and l stand for the upper and lower tails used of the distribution of the augumenting statistics.

In order to examine whether the statistics T_3 and T_4 provide additional information we compare the power of the test procedures listed in Table 1 with the power of an equivalent test (i. e., having the same level of significance under the null hypothesis) based on T_2 alone. For instance the test based on T_2 equivalent to T_{23} is $T_2 \leq p_2'$ where p_2' is such that

$$P\big(T_2 \leq p_2' \mid H_0 : N(\mu, \sigma)\big) = P\big(T_2 \leq p_2 \text{ or } \lvert T_2 \rvert \geq p_3' \mid H_0 : N(\mu, \sigma)\big)\,. \quad (2.7)$$

All the probabilities under the null and non-null hypotheses are obtained empirically using the random samples supplied by the Bell Telephone Laboratories.

The following critical values of p_2, p_3 and p_4 (constants in Table 1) are used for different values of n, the sample size. These values correspond to nearly 5% critical values for individual statistics T_2, T_3 ann T_4.

	$n = 10$	$n = 20$	$n = 30$
p_2	.9176	.9513	.9628
p_3	1.645	1.645	1.645
p_3'	1.960	1.960	1.960
p_4	1.560	1.560	1.645
p_4'	1.840	1.840	1.960

Tables 2, 3 and 4 give the empirical powers of the various test procedures given in Table 1 and of the equivalent test based on T_2 alone, along with the level of significance (L. S.) in each case attained under the null hypothesis.

Table 2
Symmetric distributions: Power of various tests compared with W (or T_2). (The figures are frequencies out of 1000.)

	Test	$T_{23(u)}$	T_2	$T_{23(l)}$	T_2	T_{23}	T_2	T_{24}	T_2	$T_{24(l)}$	T_2	$T_{24(u)}$	T_2
$n = 10$	L. S.	(.085)		(.087)		(.069)		(.088)		(.079)		(.091)	
	T (1)	606	630	618	631	600	618	684	633	580	627	714	635
	T (2)	359	372	363	372	350	354	428	374	333	364	470	364
	T (4)	184	177	174	180	162	166	191	181	155	174	234	182
	T (10)	111	101	107	102	92	92	113	104	90	98	139	104
	BE (1, 1)	97	163	100	167	95	141	141	168	227	156	91	170
$n = 20$	L. S.	(.089)		(.087)		(.075)		(.083)		(.091)		(.088)	
	T (1)	876	889	878	889	873	882	917	888	870	891	932	889
	T (2)	587	604	588	604	582	590	676	602	563	604	712	604
	T (4)	285	281	278	279	261	262	330	277	244	281	372	282
	T (10)	134	140	140	140	121	126	179	137	116	141	193	140
	BE (1, 1)	216	317	212	315	215	272	367	314	509	318	210	317
$n = 30$	L. S.	(.074)		(.076)		(.054)		(.076)		(.096)		(.076)	
	T (1)	956	972	962	972	956	962	974	972	956	974	980	972
	T (2)	728	754	742	762	728	730	808	762	716	780	830	762
	T (4)	330	342	338	350	328	302	396	350	292	374	448	350
	T (10)	144	150	142	158	142	116	192	159	126	178	208	158
	BE (1, 1)	316	452	316	400	314	354	570	474	734	542	312	474

Table 3
Skew Distributions (Figures as in Table 2).

	Test	$T_{23(u)}$	T_2	$T_{23(l)}$	T_2	T_{23}	T_2	T_{24}	T_2	$T_{24(l)}$	T_2	$T_{24(u)}$	T_2
$n = 10$	L. S.	(.085)		(.087)		(.069)		(.088)		(.079)		(.091)	
	LN (0, 1)	724	681	602	683	650	653	617	684	630	670	610	686
	L (0, 2)	118	110	98	111	98	96	123	112	101	103	147	113
	WE (.5, 1)	936	925	880	925	909	913	884	926	887	918	881	928
	WE (2, 1)	150	122	69	127	94	100	106	123	125	114	108	126
$n = 20$	L. S.	(.089)		(.087)		(.075)		(.083)		(.091)		(.088)	
	LN (0, 1)	963	949	929	948	946	945	932	948	936	949	929	948
	L (0, 2)	169	165	168	162	156	150	209	162	142	167	232	165
	WE (.5, 1)	*	*	998	*	999	*	998	*	998	*	998	*
	WE (2, 1)	238	198	148	197	198	185	158	197	199	202	172	198
$n = 30$	L. S.	(.074)		(.076)		(.054)		(.076)		(.096)		(.076)	
	LN (0, 1)	994	988	986	990	990	998	998	990	988	992	986	990
	L (0, 2)	170	196	178	208	168	144	206	208	124	220	258	208
	WE (.5, 1)	*	*	*	*	*	*	*	*	*	*	*	*
	WE (2, 1)	394	276	200	288	376	218	220	288	240	332	204	288

* Observed value is 1000.

Table 4
Discrete Distributions (Figures as in Table 2).

	Test	$T_{23(u)}$	T_2	$T_{23(l)}$	T_2	T_{23}	T_2	T_{24}	T_2	$T_{24(l)}$	T_2	$T_{24(u)}$	T_2
$n = 10$	L. S.	(.085)		(.087)		(.069)		(.088)		(.079)		(.091)	
	P (1)	746	800	721	800	725	763	722	800	747	793	720	800
	P (4)	183	203	156	205	142	180	165	205	170	193	167	209
	P (10)	137	146	106	152	106	134	140	152	133	139	142	156
$n = 20$	L. S.	(.089)		(.087)		(.075)		(.083)		(.091)		(.088)	
	P (1)	997	999	997	999	997	998	997	998	997	999	997	999
	P (4)	284	315	232	314	242	287	253	310	257	315	251	315
	P (10)	192	167	130	166	152	149	152	162	151	170	173	167
$n = 30$	L. S.	(.074)		(.076)		(.054)		(.076)		(.096)		(.076)	
	P (1)	*	*	*	*	*	*	*	*	*	*	*	*
	P (4)	372	432	314	460	324	356	322	460	342	522	320	460
	P (10)	218	214	140	220	158	164	164	220	166	260	178	220

* Observed value is 1000.

The symbols $T(1)$, $T(2)$, . . . etc. for the alternate distributions have the same meaning as in the paper by Shapiro, Wilk and Chen [8]. The numbers reported in Tables 2 to 4 are frequencies of rejecting the null hypothesis out of 1000 samples for $n = 10$ and $n = 20$ and double the observed frequencies out of 500 samples for $n = 30$.

Notes on Tables 2, 3 and 4

1. *Symmetric distributions.* Table 2 gives the power of the different test procedures compared to the equivalent Shapiro-Wilk test when the alternatives are symmetric [T(1), T(2), T(4), T(10) and BE (1, 1)].

a) Columns (5) and (6) of Table 2 show that T_{23} has nearly the same power as T_2, if not slightly inferior in some cases, which is as it should be, since T_3 is a measure of departure from symmetry.

b) Columns (7) and (8) show that T_{24} performs better than T_2, indicating that the test for normal kurtosis T_4 provides additional information.

c) Column (9) shows that the uniform distribution BE (1, 1) which is platokurtic is detected better by $T_{24(l)}$ and column (10) shows that the student distributions which are less flat are detected better by $T_{24(u)}$.

2. *Asymmetric distributions.* Table 3 gives the power of the different test procedures compared to the equivalent Shapiro-Wilk test when the alternatives are skew [LN (0, 1), L (0, 2), WE (.5, 1) and WE (2, 1)].

a) Columns (5) and (6) show that T_{23} performs better than T_2, indicating that additional information is provided by T_3.

b) Columns (6) and (7) show that T_4 is not relevant, as it should be, to test departures from symmetry.

c) Column (1) shows that distributions which are skew to the right are better detected by $T_{23(u)}$.

3. Discrete distributions

Table 4 which is similar to Tables 2 and 3 refers to the discrete distributions [P(1), P(4) and P(10)].

In these cases, we do not find any direct evidence as provided by the power functions about the additional information due to T_3 and T_4. It is conceivable that when the evidence contained jointly in T_3 and T_4 is considered, the advantage of augumenting T_2 by T_3 and T_4 may be demonstrable. It is also worth while examining whether one has to consider coefficients beyond γ_4 in the polynomial (1.2) to detect departure from normality, in addition to that provided by γ_2, in the case of the Poisson distributions.

4. Distributions of T_2 and T_3

The percentage points of T_2 and T_3 as determined empirically for $n = 10$ and 20, based on 1000 samples, and for $n = 30$ based on 500 samples are given in Table 5.

Table 5
Percentage points of T_3 and T_4, and the standard normal variable.

Prob.	T_3			T_4			$N(0, 1)$
	$n = 10$	$n = 20$	$n = 30$	$n = 10$	$n = 20$	$n = 30$	
.010	− 2.1588	− 2.3149	− 2.2778	− 1.9354	− 1.9621	− 2.1257	− 2.330
.025	− 1.8908	− 1.9406	− 1.8465	− 1.6799	− 1.7824	− 1.8266	− 1.960
.050	− 1.6586	− 1.6647	− 1.6308	− 1.4846	− 1.5771	− 1.7014	− 1.645
.100	− 1.3009	− 1.3173	− 1.3100	− 1.2918	− 1.3460	− 1.3714	− 1.281
.500	− 0.0269	− 0.0023	− 0.0127	− 0.1070	− 0.0769	− 0.1129	0
.900	1.4098	1.4095	1.3500	1.1310	1.2271	1.2332	1.281
.950	1.7115	1.6945	1.7262	1.6256	1.5497	1.6128	1.645
.975	2.0468	1.9682	1.9680	2.0217	1.8770	2.1295	1.960
.990	2.2734	2.1945	2.3770	2.3883	2.3262	2.6210	2.330

The distribution of T_3 is symmetrical and nearly normal for the values of n studied; the distribution of T_4 appears to be somewhat skew; however, the percentage points over 50% agree closely with the normal values, while the lower percentage points are shifted somewhat to the right.

5. Conclusions

5.1 Previous studies supplemented by our own indicate that the Shapiro-Wilk statistic provides an omnibus test for normality and its performance is generally better than other known tests for normality.

5.2 Shapiro-Wilk test does not by itself indicate whether the detected departure from normality is due to skewness, non-normal kurtosis etc. We, therefore, propose to supplement the W-test by computing other statistics and examining their magnitudes. A natural way of doing this is to fit a polynomial to the observed order statistics using expected values of order statistics under the normal hypothesis as abscissa. The coefficient of the first degree term in the fitted polynomial provides the Shapiro-Wilk statistic, W. The coefficient of the second degree term represents the deviation from symmetry, and the significance of skewness is examined by the statistic T_3, using the percentage points given in Table 5 if necessary. Large positive values of T_3 indicate skewness of the right tail and large negative values, skewness of the left tail. It appears from our study that an examination of T_3, while indicating the nature of departure, also provides *additional information* independently of W in detecting deviation from normality. Thus the use of W and T_3 leads to a more efficient test procedure for normality when there is departure from symmetry.

5.3 The coefficient of the third degree term measures departure from normal kurtosis, and its significance is tested by using the statistic T_4 in the same way as T_3 for detecting asymmetry. Our study indicates that T_4 carries additional information on deviation from normality independently of W when there is departure from normal kurtosis.

5.4 We have not considered test procedures which use the three statistics T_2, T_3 and T_4 simultaneously or considered tests based on coefficients of higher degree terms in the fitted polynomial. It is possible that such tests would be more efficient than W alone, in the case of alternatives like the Poisson distribution.

5.5 It would be of some interest to examine whether augumenting the W test by tests based on estimates of Pearson's β_1 and β_2 instead of T_3 and T_4 would provide more efficient test procedures. However, as observed earlier, the choice of T_3 and T_4 arise in a natural way in examining the relationship between observed order statistics and the expected values under the normal hypothesis. It may also be of some interest to consider other ways of augumenting the W test by T_3 and T_4, which may be more efficient than those described in the paper.

5.6 In the definitions of statistics (2.4) to (2.6) one could use the identity matrix instead of b.f. as suggested by Shapiro and Francia [10] for the W statistic and make an investigation similar to ours. In cases where the covariances of order statistics are not available one could use a diagonal matrix representing the variances of order statistics in the place of b.f.

5.7 In a subsequent communication, we propose to discuss the asymptotic power properties of the test procedures given by us to augument the W-test.

The main object of our investigation was to demonstrate the existence of tests more powerful than the W test of Shapiro-Wilk for specific types of alternatives. These are obtained by augumenting the W test with other tests indicating departure from symmetry and normal kurtosis. It has not been possible for us to consider different ways of combining these individual tests or provide combined test criteria for given levels of significance such as 5 percent, 1 percent etc. We have adopted the simple approach of carrying out individual tests at the 5 percent level and rejecting the null hypothesis if at least one test shows significance, which does not provide 5 percent significance level for the combined test. Our investigation has shown that the W test is not uniformly the best under all alternatives and there exist other tests which are more powerful under specific alternatives. Further research in this direction would be useful.

6. Acknowledgments

We wish to thank R. Gnanadesikan who has kindly supplied us with tapes of samples from a number of distributions available at the Bell Telephone Laboratories, which made our investigation possible. Wayne Johnson wrote the computer programs for the various tests. We admire his patience in working with us and thank him for the care with which he has handled the computations.

References

[1] DURBIN, J. and KNOTT, M. (1972): Components of Cramer-von Mises statistics. 1. J. Roy, Statist. Soc. *B 34*, 290–307.

[2] HECKARD, R. (1970): Master's thesis. Pennsylvania State University.

[3] LINDER, A. and CZEGLEDY, P. (1973): Normality test by fractile graphical analysis. Sankhya....

[4] MAHALANOBIS, P. C. (1960): A method of fractile graphical analysis. Econometrica *28*, 325–351.

[5] NEYMAN, J. (1937): 'Smooth test' for goodness of fit. Skand. Aktuartidskr. *20*, 149–199.

[6] RAO, C. R. (1973): Linear Statistical Inference and its Applications (second Edition). John Wiley & Sons.

[7] RAO, C. R., MATTHAI, A. and MITRA, S. K. (1966): Formulae and Tables for Statistical Work.

[8] SHAPIRO, S. S., WILK, M. B. and CHEN, H. J. (1968): A comparative study of various tests for normality. J. Amer. Statist. Assn. *63*, 1343–1372.

[9] SHAPIRO, S. S. and WILK, M. B. (1965): An analysis of variance test for normality (complete samples). Biometrika *52*, 591–611.

[10] SHAPIRO, S. S. and FRANCIA, R. S. (1972): An approximate analysis of variance test for normality. J. Amer. Statist. Assn. *67*, 215–216.

[11] STEPHENS, M. A. (1973): EDF statistics for goodness-of-fit and some comparisons. Tech. Report, McMaster University.

Addresses of the authors:
M. L. Puri, Indiana University.
C. Radhakrishna Rao, Indiana University and Indian Statistical Institute.

Klaus Abt

Klaus Abt

Fitting Constants in Cross-Classification Models with Irregular Patterns of Empty Cells

Summary

A method is proposed for fitting main effect and interaction constants in analysis of variance models for non-orthogonal cross-classified data layouts with irregular patterns of empty cells. The model is assumed to be of the fixed-effects type with all factors having qualitative levels. The confounding that possibly exists among the effects (caused by the presence of empty cells) is treated by using the concept of 'cell-identities' which is defined. The testability of null hypotheses on the fitted constants is demonstrated. A numerical example is given to illustrate the analysis of models consisting of constants fitted according to the proposed method. The application of the method may serve any or all of the following purposes: a) to screen a given layout of incomplete and unbalanced data for significant factorial effects, b) to identify, if present, confounded effects and c) to yield an estimate of the experimental error if otherwise not obtainable.

1. **Introduction**

The classical method of fitting constants for the analysis of variance of unbalanced data layouts as introduced by Yates [11] concerns the fitting of main effect constants (parameters) only. Except for the interaction in the case of the two-way cross classification where the interaction sum of squares equals the lack of fit, null hypotheses about main effects only are tested assuming zero interactions.

Comparatively little work has been done to extend the method of fitting constants to the case in which the interactions cannot be ignored. For example, in the procedure outlined by Bock [3], the additional constants which can be fitted after all main effect constants have been fitted are somewhat arbitrarily designated as interaction constants. Essentially, Bock's method for the non-orthogonal ANOVA is based upon the assumption that the analyst knows which constants he can fit, and no procedure is given for determining the constants that can be fitted.

Other authors have taken different approaches to the general problem of analysis of variance for unbalanced data layouts when interactions are assumed present. Federer [6] has given an extensive account of the methods of analysis available for this situation. Elston and Bush [5] thoroughly investigated the possibilities of testing hypotheses about the main effects when the model contains interactions. Gorman and Toman [7] and Dykstra [4] have treated the analysis of incomplete and unbalanced data classifications when all factors are quantitative, i.e., when the levels of all factors are values of a variate scale.

Apart from the problems connected with the analysis of variance (such as the proper assignment of sums of squares to factorial effects), the main difficulty with methods of fitting constants in interaction models for unbalanced data layouts stems from the presence of empty cells, i.e., from cells, as defined by factor level combinations, for which there are no observations available. Once all cells are occupied, the full set of main effect constants and interaction constants can be fitted.

In the present paper a method is proposed for determining the constants which can be fitted, in a fixed effects interaction model with qualitative factor levels only, when there is an arbitrary number of empty cells. The irregular pattern of empty cells is assumed to have originated from lost data and/or from the impossibility of obtaining values for the factor level combinations concerned. The method is exemplified with two-way and three-way cross classifications. Generalizations of the method for the general n-way cross classification, for partially nested classifications and for situations involving quantitative factors are not difficult.

As Federer [6] emphasized, any analysis of variance for unbalanced data classifications with empty cells is biased simply because some of the parameters in the interaction model have to be equated to zero. In full recognition of this fact, the proposed method is based on the assumption that the analyst wants to utilize all information available in the given data, however incomplete the data may be. Using the method, the analyst may be able to make inferences about factorial effects based upon the incomplete sets of degrees of freedom available for testing hypotheses concerning the said effects. After all, a rejected null hypothesis, although based upon an incomplete set of degrees of freedom, will hint at the importance of the factorial effect concerned.

Making use of the available data in the indicated way constitutes a case of model building and hypothesis testing according to the available data, in the present situation particularly according to the irregular pattern of non-empty cells. Being aware of this fact while applying the method, the analyst should not expect to be able to more than screen the available data for 'significant' factorial effects. The use of such a screening method follows the tendency of accepting 'datadredging procedures' when the user is aware of the limitations, as discussed by Selvin and Stuart [10]. Therefore, the method of fitting constants in case of empty cells as it will be outlined in Sections 3 and 4 is mainly intended for screening purposes. The emphasis is put on hypothesis testing; the problem of obtaining unbiased estimates of factorial effects is not discussed.

2. Basis of the Analysis

The procedure of fitting constants to be described cannot be separated from the specific method of non-orthogonal analysis of variance which will be applied to the linear model containing the fitted constants. Such a method of non-orthogonal ANOVA was proposed by the present author (Abt [1]). For convenience, this method will briefly be outlined in the following. Basis of the analysis is the general linear model of the form

$$y = Y + e = \mu + \sum_{\nu=1}^{N} k_\nu \, x_\nu + e \,, \tag{1}$$

where the residual terms e are $NID(0, \sigma^2)$ and the k_ν, $\nu = 1, \ldots, N$ are the model constants representing the factorial effects, μ being the general constant. The x_ν are auxiliary variables taking the values 0 or 1 only.

The non-orthogonal ANOVA starts from a model of full rank N containing the constants k_ν, $\nu = 1, \ldots, N$, where N is the sum of the degrees of freedom of all factorial effects to be analyzed. In order to make the model one of full rank, the linear restrictions of Graybill [8] are used. For example, in a two-way cross classification assumed 'complete', i.e. having no empty cells, with factors $\mathscr{A}$ and $\mathscr{B}$, factor levels $\alpha = 1, 2, \ldots, A$ and $\beta = 1, 2, \ldots, B$, respectively, main effect constants a_α and b_β, and interaction constants $ab_{\alpha\beta}$, the Graybill restrictions are:

$$a_A = b_B = a\,b_{\alpha B} = a\,b_{A\beta} = 0 \begin{cases} \alpha = 1, 2, \ldots, A \\ \beta = 1, 2, \ldots, B\,. \end{cases} \tag{2}$$

These restrictions leave all factorial effects as contrasts with respect to the last factor levels, e.g., $a_\alpha - a_A = a_\alpha$. As will be seen, the Graybill restrictions are very convenient for the proposed procedure of fitting constants. Using the restrictions, model (1) can be written for cell (α^*, β^*), in case of the complete two-way cross classification, as follows:

$$\begin{aligned} y_{\alpha^*\beta^*} &= Y_{\alpha^*\beta^*} + e_{\alpha^*\beta^*} \\ &= \mu + \sum_{\alpha=1}^{A-1} a_\alpha \, x_\alpha^{(\mathscr{A})} + \sum_{\beta=1}^{B-1} b_\beta \, x_\beta^{(\mathscr{B})} + \sum_{\alpha=1}^{A-1} \sum_{\beta=1}^{B-1} ab_{\alpha\beta} \, x_{\alpha\beta}^{(\mathscr{A}\mathscr{B})} + e_{\alpha^*\beta^*} \,. \end{aligned} \tag{3}$$

In this notation, the auxiliary variables x take the values:

$$x_{\alpha}^{(\mathscr{A})} = \begin{cases} 1 & \text{if} \quad \alpha = \alpha^* \\ 0 & \text{if} \quad \alpha \neq \alpha^* , \end{cases}$$

$$x_{\beta}^{(\mathscr{B})} = \begin{cases} 1 & \text{if} \quad \beta = \beta^* \\ 0 & \text{if} \quad \beta \neq \beta^* , \end{cases}$$

$$x_{\alpha\beta}^{(\mathscr{A}\mathscr{B})} = \begin{cases} 1 & \text{if} \quad \alpha, \beta = \alpha^*, \beta^* \\ 0 & \text{if} \quad \alpha, \beta \neq \alpha^*, \beta^* . \end{cases}$$

In the non-orthogonal ANOVA, the linear model of form (3) is evaluated by a step-down-procedure. Evaluating the general linear model step-down-wise rather than in a build-up-manner is mandatory because of the possible existence of a 'compound' of the independent (here: auxiliary) variables x_{ν}, or of several such compounds among the N variables x_{ν}. The definition of a compound is as follows: $\tilde{N} \leq N$ independent variables x_{ν} form a compound with estimated residual variance $\hat{\sigma}^2_{\tilde{N}}$ when the residual variance increases by orders of magnitude to $\hat{\sigma}^2_{\tilde{N}-1}$ if any one variable $x_{..}$ is deleted from the compound: $\hat{\sigma}^2_{\tilde{N}-1} \gg \hat{\sigma}^2_{\tilde{N}}$. In other words, all $\tilde{N}$ independent variables of a compound are needed in the model to provide for a small residual variance. In a step-down-procedure according to one of the usual prediction-power-for-y criteria such a compound will be left complete until the latest possible step. In contrast to this, in a build-up-procedure, the last of the $\tilde{N}$ variables of a compound may be incorporated into the model only with the last step. Numerical examples of compounds are given in Abt [1].

The step-down-procedure proposed is based on a cumulative ranking of the factorial effects by the following criterion:

$$I(X) = \int_{F_c}^{\infty} \varphi(F)\, dF \tag{4}$$

where $\varphi(F)$ is the density function of the F-distribution, and $I(X)$ is the incomplete Beta-function with vector of arguments $X = (1 + f_1 f_2^{-1} F_c)^{-1}, \frac{1}{2} f_2, \frac{1}{2} f_1$. F_c is the calculated F-value at a given step of the procedure:

$$F_c = \frac{\dfrac{SSR_N - SSR_{N'}}{N - N'}}{\dfrac{SSE_N}{n - N - 1}}, \tag{5}$$

with degrees of freedom $f_1 = N - N'$ and $f_2 = n - N - 1$.

The symbols in (5) have the meaning:

SSR_N = Sum of Squares due to Regression of all N auxiliary variables x_ν in original model;
$SSR_{N'}$ = Sum of Squares due to Regression of $N' < N\, x_\nu$, where $N - N'$ equals the sum of degrees of freedom of the factorial effects contained in the contribution $SSR_N - SSR_{N'}$;
SSE_N = Sum of Squares due to Error in original model;
n = Total number of observations y.

The choice of (4) as ranking criterion appears suitable because $I(X)$ yields a measure of the prediction power for y for each one of a group of competing factorial effects taking into account their degrees of freedom $N - N'$. A criterion such as $SSR_N - SSR_{N'}$, for instance, would lack this advantage.

At the first step of the ranking procedure, that factorial effect [represented by a group of constants k_ν in (1)] is considered least important for y which among the 'admissible' effects yields $\operatorname{Max} I(X)$. (The concept of admissibility is explained below.)

At the second step, the factorial effect found least important in the first step is combined with each effect, one by another, which has become admissible by the deletion of the first effect, to find that pair of effects which yields $\operatorname{Max} I(X)$. The new effect which leads to this pair is ranked second to least important of all effects. The procedure continues until all effects have been ranked. It is extremely important to notice that all contributions $SSR_N - SSR_{N'}$ to the regression sum of squares must be understood as contributions to SSR_N in addition to the contributions of the effects contained in the model of the N' independent variables. The user of the method may choose a level of $I(X)$ which, when reached the first time at a given step of the procedure, is to determine the 'significant model'.

The concept of admissibility of the effects for the ranking process is as follows. According to Scheffé [9], the contribution $SSR_N - SSR_{N'}$ is dependent upon the linear restrictions chosen if in the remaining model of the N' variables there is a factorial effect which represents an interaction effect with an effect contained in $SSR_N - SSR_{N'}$. For example, in the two-way cross classification discussed above, the contribution of the main effect $\mathscr{A}$ is dependent upon the linear restrictions chosen as long as the interaction effect $\mathscr{AB}$ is contained in the remaining model of N' variables. However, since in general the linear restrictions are arbitrary—like the Graybill restrictions chosen in the method—it is desirable to make the contributions $SSR_N - SSR_{N'}$ independent from the choice of the linear restrictions. Therefore, in the ranking process described before, a factorial effect is admissible for ranking at a given step only if no effect remains in the model of the N' variables which represents an interaction effect with the effect to be ranked. In the above example, at the first step the only effect admissible for ranking is the interaction $\mathscr{AB}$. The admissibility restriction is not severe for the ranking process since, in analysis of variance, the presence of significant interaction effects causes inferences about the constituent lower order effects of the significant interactions to be of secondary importance.

3. Proposed Method of Fitting Constants

The method of fitting constants will be introduced using specific examples of two-way and three-way cross classifications. First, the basic principles of the method are discussed with a 3×4 classification having unequal cell frequencies $n_{\alpha\beta} > 0$, as shown in Fig. 1. The assumption of unequal $n_{\alpha\beta} > 0$ implies at least one $n_{\alpha\beta} > 1$ such that a within-cell estimate of the experimental error can be obtained. An '$\times$' stands for an occupied cell, i.e., in Example 1 all 12 cells are marked by an '$\times$'.

		$\mathscr{B}$					
		$\beta = 1$	$\beta = 2$	$\beta = 3$	$\beta = 4$		
	$\alpha = 1$	$\otimes$	$\otimes$	$\otimes$	$\times$	$\otimes$	(1, .)
$\mathscr{A}$	$\alpha = 2$	$\otimes$	$\otimes$	$\otimes$	$\times$	$\otimes$	(2, .)
	$\alpha = 3$	$\times$	$\times$	$\times$	$\times$	$\times$	(3, .)
		$\otimes$	$\otimes$	$\otimes$	$\times$		
		(., 1)	(., 2)	(., 3)	(., 4)		

Fig. 1
Data layout of Example 1.

Corresponding to the step-down ranking procedure for the factorial effects (Section 2), the fitting of the $\mathscr{A}\mathscr{B}$-interaction constants is discussed first. Fig. 1 shows, besides the 3×4 layout of the original cells (α, β), the marginal rows $(\alpha, .)$ and $(., \beta)$ of cells, using an obvious notation. Each constant to be fitted will be 'based' upon an original cell or a marginal cell, i.e., the constant will have the index α and/or β of the cell upon which it is based. Using the Graybill restrictions (2) it is obvious that no interaction constants can be fitted based upon the last rows $\alpha = A = 3$ and $\beta = B = 4$. This leaves $2 \times 3 = 6$ cells upon which interaction constants may be based: these 6 cells are marked by circles around the '$\times$''s in Fig. 1. The corresponding constants fitted are ab_{11}, ab_{12}, ab_{13}, ab_{21}, ab_{22} and ab_{23}.

Each first-order-interaction constant fitted must be interpretable as a contrast of contrasts: such a constant is to measure the difference between a) the difference of two levels of the first factor at a given level of the second factor and b) the difference of the same two levels of the first factor at another level of the second factor. A contrast can be subjected to a null hypothesis test if a linear combination $L(y)$ of observations $y_{\alpha\beta}$ exists whose expectation is the contrast under the given model, see for example, Graybill [8]. In other words 'testability' implies that the constant to be fitted must be estimable. A sufficient condition for the estimability of interaction constants is that for each such constant fitted there is a rectangle of four occupied cells in the two-way cross classification, where one corner of the rectangle is the cell upon which the constant is based and where the other three corner-cells are not bases of interaction constants fitted. Such a rectangle will be called a 'quadrupel' of cells. At least as many interaction constants of first order can be fitted as there are different quadrupels of cells in the layout. One such quadrupel is the one formed by the 4 cells (1, 1), (1, 4), (3, 1), (3, 4). A linear combination $L(y)$ based upon the observations in these cells is as follows: $L(y) = (y_{11} - y_{14}) - (y_{31} - y_{34})$. Under model (3), with $A = 3$ and $B = 4$, $L(y)$ has expectation:

$$E[L(y)] = (\mu + a_1 + b_1 + ab_{11}) - (\mu + a_1) - (\mu + b_1) + \mu = ab_{11} .$$

Hence, the interaction constant ab_{11} is estimable and can be subjected to the null hypothesis $ab_{11} = 0$, i.e., ab_{11} can be included in the model.

The other five quadrupels and their corresponding $L(y)$ combinations are formed correspondingly, where the $E[L(y)]$ are the remaining 5 interaction constants which can be fitted: ab_{12}, ab_{13}, ab_{21}, ab_{22}, ab_{23}, respectively. Since all 6 constants are estimable under the given model, they can be subjected to the joint null hypothesis for $\mathscr{AB}$: $ab_{11} = ab_{12} = ab_{13} = ab_{21} = ab_{22} = ab_{23} = 0$.

To complete the discussion of Example 1, only main effect constants remain to be fitted. Under the Graybill restrictions, each main effect constant represents a contrast of the effects of two levels of a factor. Again, testability of such a contrast requires that there is a linear combination $L(y)$ of observations with the constant to be fitted as expectation under the given model. For the constant a_1, for instance, one has $E[L(y)] = E[y_{14} - y_{34}] = \mu + a_1 - \mu = a_1$. In a corresponding manner, the other 4 main effect constants a_2, b_1, b_2 and b_3 can be shown estimable. The 5 constants are based upon the 5 marginal cells as indicated by circles in Fig. 1. Altogether, therefore, the layout of Example 1 allows 11 constants to be fitted: 5 main effect constants and 6 interaction constants. These 11 constants correspond to the 11 degrees of freedom 'between (occupied) cells'.

Example 2 (Fig. 2) is derived from Example 1 by deleting all observations in cells (1, 1), (1, 4), (2, 1) and (2, 3). In these 4 cells, therefore, one now has $n_{\alpha\beta} = 0$.

		$\mathscr{B}$			
		$\beta = 1$	$\beta = 2$	$\beta = 3$	$\beta = 4$
	$\alpha = 1$		⊗	×	
$\mathscr{A}$	$\alpha = 2$		⊗		×
	$\alpha = 3$	×	×	×	×

Fig. 2
Data layout of Example 2.

Again starting with the fitting of interaction constants, there are 2 different quadrupels of cells upon which to base the 2 constants ab_{12} and ab_{22}, respectively. These are the only estimable interaction constants as can easily be seen. All 5 main effect constants can be estimated under the given model containing 7 constants. The 7 constants correspond to the 7 degrees of freedom between the 8 occupied cells.

Example 3 (Fig. 3) is a 4×4 classification with 9 occupied and 7 empty cells:

$\mathscr{A}$ \ $\mathscr{B}$	$\beta = 1$	$\beta = 2$	$\beta = 3$	$\beta = 4$
$\alpha = 1$		⊗	×	
$\alpha = 2$			⊗	×
$\alpha = 3$	×	×	×	
$\alpha = 4$		×		×

Fig. 3
Data layout of Example 3.

One interaction constant, namely ab_{12}, can be based upon the quadrupel (1, 2), (1, 3), (3, 2), (3, 3). Assuming, for the moment, that $3 + 3 = 6$ main effect constants can be fitted (which is to be shown), there is one of the 8 degrees of freedom between the 9 occupied cells left for a second interaction constant. But there is no second quadrupel upon which to base another interaction constant. Nevertheless, a second interaction constant can be based, for example, upon cell (2, 3), that is, ab_{23}, under the given model then containing a_1, a_2, a_3, b_1, b_2, b_3, ab_{12} and ab_{23}:

$$E[\{(y_{23} - y_{24}) + (y_{32} - y_{33})\} - (y_{42} - y_{44})] = ab_{23} .$$

The possibility of combining—in this example—rows $\alpha = 2$ and $\alpha = 3$ to yield an estimate of ab_{23} shows that the existence of quadrupels is not a necessary condition for the fitting of interaction constants. All 6 main effect constants can be included in the model and subjected to null hypotheses because each constant is estimable under the given model with 8 constants:

$$E[(y_{13} - y_{33}) + (y_{32} - y_{42})] = (\mu + a_1 + b_3) - (\mu + a_3 + b_3)$$
$$+ (\mu + a_3 + b_2) - (\mu + b_2) = a_1 .$$

Therefore, also a main effect constant (here a_1) may be fitted only because it is possible to combine 2 rows to yield an estimate of that constant. The other 5 main effect constants follow immediately: $E[y_{24} - y_{44}] = a_2$, $E[y_{32} - y_{42}] = a_3$, $E[(y_{31} - y_{32}) + (y_{42} - y_{44})] = b_1$, $E[y_{42} - y_{44}] = b_2$, $E[(y_{33} - y_{32}) - (y_{44} - y_{42})] = b_3$.

Notice that constant b_3 is shown to be estimable by using, in $L(y)$, two differences of observations with respect to column $\beta = 2$. Summarizing for Example 3, there are indeed eight constants which can be fitted, namely $3 + 3 = 6$ main effect constants and the two interaction constants ab_{12} and ab_{23}.

Example 4 (Fig. 4) is a complete 3-way classification with factors $\mathscr{A}, \mathscr{B}, \mathscr{C}$ having 3, 3, 2 levels, respectively. The example is intended to show the fitting of second order interaction effects only; the fitting of the first order interaction and main effects easily follows the procedure shown before.

			$\mathscr{B}$		
		$\mathscr{C}$	$\beta = 1$	$\beta = 2$	$\beta = 3$
$\mathscr{A}$	$\alpha = 1$	$\gamma = 1$	⊗	⊗	×
		$\gamma = 2$	×	×	×
	$\alpha = 2$	$\gamma = 1$	⊗	⊗	×
		$\gamma = 2$	×	×	×
	$\alpha = 3$	$\gamma = 1$	×	×	×
		$\gamma = 2$	×	×	×

Fig. 4
Data layout of Example 4.

An interaction constant of second order is to measure the change of a first-order interaction effect (between 2 of the 3 factors) when going from one level to another level of the third factor. Accordingly, a sufficient condition for fitting a second-order interaction constant is that there exists an 'octupel' of occupied cells being composed of two corresponding quadrupels at two levels of the third factor. Such a constant can be subjected to a null hypothesis since for the octupel upon which the constant is based a linear combination $L(y)$ can be defined the expectation of which is the constant to be fitted under the given model. For example, abc_{111} is based upon cell (1, 1, 1) and can be shown estimable under the model containing a_1, a_2, b_1, b_2, c_1, ab_{11}, ab_{12}, ab_{21}, ab_{22}, ac_{11}, ac_{21}, bc_{11}, bc_{21}, abc_{111}, abc_{121}, abc_{211}, abc_{221}:

$$E[L(y)] = E[\{(y_{111} - y_{131}) - (y_{112} - y_{132})\} \\ - \{(y_{311} - y_{331}) - (y_{312} - y_{332})\}] = abc_{111} .$$

The four cells corresponding to the different octupels upon which the four abc-constants can be based are indicated by circles in Fig. 4.

From the examples discussed so far the fitting procedure can be seen to enable the analyst to obtain an estimate of the experimental error in cross classifications with 3 or more factors when a within-cell estimate is not available. In a situation like this, usually the highest order interaction is taken as basis to estimate σ^2. In the types of data layouts discussed, separating the highest order interaction from the factorial effects of lower order appears possible only by using a method of fitting constants like the one proposed.

4. Fitting of Constants with Confounding Present

The fitting process becomes more complicated when—in the case of empty cells—confounding is present among the factorial effects. In the present context, such confounding is assumed to occur accidentally. This assumption is in accordance with the presumed irregular occurrence of empty cells due to lost values or impossible factor level combinations.

Example 5 (Fig. 5) shows a 3×3 classification with 5 of the 9 cells occupied:

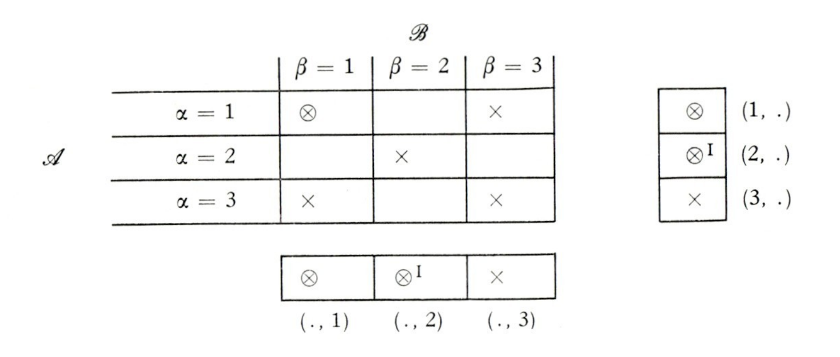

Fig. 5
Data layout of Example 5.

There are 4 degrees of freedom between occupied cells of which one degree of freedom seemingly can be assigned to an interaction constant, ab_{11}, see the circle in cell (1, 1). There remain three degrees of freedom for main effect constants, although, inspecting the marginal cells $(\alpha, .)$ and $(., \beta)$, four main effect constants appear as if they can be fitted. The apparent possibility of fitting four main effect constants is due to the fact that cell (2, 2) is occupied, i.e., both marginal cells (2, .) and (., 2) appear occupied only because of cell (2, 2) being occupied. If such a relation between two marginal cells exists it will be called a 'cell-identity', symbolized in the present case by $(2, .) \equiv (., 2)$. The cells involved in a cell identity will be marked by an 'I', see Fig. 5.

A cell identity represents 1 confounded degree of freedom. In the present case it represents the confounding between the two main effects of factors $\mathscr{A}$ and $\mathscr{B}$. Three linear combinations of observations can be constructed to yield estimates of the main effect constants under the model containing a_1, a_2, b_1, b_2, ab_{11}: $E[y_{13} - y_{33}] = a_1$, $E[y_{22} - y_{33}] = a_2 + b_2$, $E[y_{31} - y_{33}] = b_1$. To eliminate confounding one may either set $a_2 = 0$ or $b_2 = 0$. This leads to two possible models, both containing $N = 4$ constants corresponding to the 4 degrees of freedom between the 5 occupied cells (using an obvious notation for the models):

$$\text{Model I:} \quad Y^{(I)} = (\mu, a_1, b_1, b_2, ab_{11}) ,$$

$$\text{Model II:} \quad Y^{(II)} = (\mu, a_1, a_2, b_1, ab_{11}) .$$

In order to analyze the data layout by the ranking method for factorial effects, one may perform one analysis for each model. The 'significant model' which needs the smallest number of degrees of freedom to explain the differences between the occupied cells may then be considered as the 'most economical' model. For example, using Model I and assuming $H_0\{ab_{11} = 0\}$ not to have been rejected, the analysis may show that the additional contribution of a_1 to SSR_N is not significant. If the main effect of $\mathscr{B}$ represented by b_1 and b_2 is significant, the 'significant model' would consist of b_1 and b_2 alone. Accordingly, the inference would be that the effects of factor $\mathscr{A}$ are not needed to explain the differences between the 5 occupied cells. On the other hand, analysis of Model II may show that the contribution of b_1 in addition to that of a_1 and a_2 is needed to explain the differences. The latter would imply that the significant model would contain the three constants a_1, a_2 and b_1. Obviously then, Model I would lead to a 'more economic' solution, although this result would not automatically imply that factor $\mathscr{A}$ is of no importance. The only possible conclusion could be that the effects of factor $\mathscr{B}$ alone are sufficient to explain the differences between the 5 occupied cells.

Example 6 (Fig. 6a and 6b) is presented to show how to deal with confounded interaction effects. The layout results from that of Example 4 by deletion of the observations in 5 cells. This example will be used as basis for a numerical example (see Section 6).

$\mathscr{A}$		$\mathscr{C}$	$\mathscr{B}$: $\beta = 1$	$\beta = 2$	$\beta = 3$
	$\alpha = 1$	$\gamma = 1$	⊗		×
		$\gamma = 2$	×		×
$\mathscr{A}$	$\alpha = 2$	$\gamma = 1$	×	×	
		$\gamma = 2$		×	×
	$\alpha = 3$	$\gamma = 1$	×		×
		$\gamma = 2$	×	×	×

Fig. 6a
Data layout of Example 6.

There is one octupel of cells upon which the second-order interaction constant abc_{111} can be based. In order to investigate the fitting of first-order interaction constants, the given data layout of Fig. 6a is folded such that the 3 two-factor classifications shown in Fig. 6b result.

		$\mathscr{B}$		
		$\beta = 1$	$\beta = 2$	$\beta = 3$
	$\alpha = 1$	$\otimes$		$\times$
$\mathscr{A}$	$\alpha = 2$	$\otimes^{I}$	$\otimes$	$\times$
	$\alpha = 3$	$\times$	$\times$	$\times$

		$\mathscr{C}$	
		$\gamma = 1$	$\gamma = 2$
	$\alpha = 1$	$\otimes$	$\times$
$\mathscr{A}$	$\alpha = 2$	$\otimes^{I}$	$\times$
	$\alpha = 3$	$\times$	$\times$

		$\mathscr{B}$		
		$\beta = 1$	$\beta = 2$	$\beta = 3$
$\mathscr{C}$	$\gamma = 1$	$\otimes$	$\otimes^{I}$	$\times$
	$\gamma = 2$	$\times$	$\times$	$\times$

Fig. 6b
Interactions $\mathscr{AB}$, $\mathscr{AC}$ and $\mathscr{BC}$ of Example 6.

Searching for quadrupels upon which to base first-order interaction constants, one finds three quadrupels for $\mathscr{AB}$-constants and two each for $\mathscr{AC}$- and $\mathscr{BC}$-constants. Assuming that all $2 + 2 + 1 = 5$ main effect constants can be fitted, the model would contain, besides μ, $1 + 3 + 2 + 2 + 5 = 13$ constants. However, there are only 12 degrees of freedom between the 13 occupied cells. Obviously, one degree of freedom is confounded, that is, there must be one cell-identity in the layout. It is not difficult to find the identity to read as follows:

$$(2, 1, .) + (., 2, 1) \equiv (2, ., 1) \ . \tag{6}$$

One way to interpret this cell-identity is that deleting all observations from the two marginal cells on the left-hand side of (6) automatically leads to cell (2, ., 1) being empty too. Hence, not all three interaction constants based on the three cells can be included in the model at the same time. Only two of the three constants ab_{21}, bc_{21} and ac_{21} may be fitted at a time.

Assuming, for the time being, the model to consist of all 13 constants, the following expectations of 13 linear combinations of observations y show which of the 13 constants may be fitted at the same time:

$$E[\{(y_{111}-y_{112})-(y_{131}-y_{132})\}-\{(y_{311}-y_{312})-(y_{331}-y_{332})\}]=abc_{111},$$

$$E[(y_{112}-y_{132})-(y_{312}-y_{332})]=ab_{11},$$

$$E[(y_{222}-y_{322})-(y_{232}-y_{332})]=ab_{22},$$

$$E[(y_{131}-y_{132})-(y_{331}-y_{332})]=ac_{11},$$

$$E[(y_{311}-y_{312})-(y_{331}-y_{332})]=bc_{11},$$

$$E[(y_{211}-y_{311})-(y_{232}-y_{332})]=ab_{21}+ac_{21},$$

$$E[(y_{221}-y_{222})-(y_{331}-y_{332})]=ac_{21}+bc_{21},$$

$$E[(y_{211}-y_{311})-\{(y_{221}-y_{222})+(y_{232}-y_{331})\}]=ab_{21}-bc_{21},$$

$$E[y_{132}-y_{332}]=a_1,$$

$$E[y_{232}-y_{332}]=a_2,$$

$$E[y_{312}-y_{332}]=b_1,$$

$$E[y_{322}-y_{332}]=b_2,$$

$$E[y_{331}-y_{332}]=c_1.$$

Three competing models may be constructed by putting either $bc_{21}=0$, or $ac_{21}=0$, or $ab_{21}=0$. Each of these three models will consist of a common model part $Y^{(0)}$ containing, besides μ, 10 constants, and additionally of one of the three pairs of first-order interaction constants involved in the identity (6), as follows:

Model I: $Y^{(\mathrm{I})}=(Y^{(0)}, ab_{21}, ac_{21})$,

Model II: $Y^{(\mathrm{II})}=(Y^{(0)}, ab_{21}, bc_{21})$,

Model III: $Y^{(\mathrm{III})}=(Y^{(0)}, ac_{21}, bc_{21})$,

where $Y^{(0)}=\mu, a_1, a_2, b_1, b_2, c_1, ab_{11}, ab_{22}, ac_{11}, bc_{11}, abc_{111}$.

As in example 5, the model leading to the smallest number of degrees of freedom to explain the differences between the 13 cell averages will be called the most economical model (see Section 6).

From the discussion in this section the proposed method of fitting constants can be seen to be of value also for the detection of confounding alone. For example, in an experiment of sample survey with difficult data acquisition the experimenter may wish to continue collecting data until all confounding has disappeared from the data layout arrived at. The decision whether or not confounding has disappeared will be possible by application of the fitting method.

5. Summary of Fitting Procedure

From the examples discussed in the two preceding sections, a set of rules can be deduced for the fitting of constants when following the proposed procedure. In accordance with the type of examples discussed, the rules will cover given data layouts of three-way and two-way cross classifications only. The treatment of more-than-three-way classifications follows the same principles as the ones demonstrated, although the fitting procedure becomes rather complicated with the number of factors in the given layout exceeding 3.

In any such specific case, nevertheless, the user of the method should be able to generalize the principles outlined for the three-way and two-way classification. The analyst, when faced with an unbalanced and incomplete more-than-three-way classification, will probably not analyze higher than second-order interaction effects which fact will ease the generalization of the method.

The rules are as follows:

5.1 Assure that in the given cross-classified data layout each level of each factor is represented by at least one observation. Delete 'empty' levels from the layout if necessary. At least one cell in the given data layout must be occupied by more than one observation such that an estimate of the experimental error 'within cells' can be obtained. In a three-way classification with only one observation in each occupied cell only first-order interaction and main effect constants can be fitted.

5.2 If the given layout is a three-way classification, fold the layout into the three possible two-way classifications (see Example 6).

5.3 Inspect, when applicable, the three-way classification for the number of different octupels of cells. Fit, based upon the upper left hand corner of each such octupel, one second-order interaction constant (see Examples 4 and 6).

5.4 Inspect the two-way classification(s) for the number of different quadrupels of cells. Fit, based upon the upper left hand corner of each such quadrupel, one first-order interaction constant (see all examples).

5.5 Fit, for each factor, all possible main effect constants, i.e., one constant based upon each level of each factor except for the last levels.

5.6 Count the number N^* of constants fitted according to rules 5.3, 5.4 and 5.5, as applicable.

  5.61 If $N^* = N$ = number of degrees of freedom between occupied cells in the given data layout, apply the ranking method for factorial effects (Section 2) to the model containing the N^* constants.

  5.62 If $N^* < N$, investigate the possibility of 'combining' rows of cells to fit the missing constant(s) (see Example 3).

  5.63 If $N^* > N$, search for confounded effects, i.e. search for cell-identities. Find, corresponding to each such identity, one confounded constant (see Examples 5 and 6).

  5.64 In both cases 5.62 and 5.63 construct linear combinations of the observations y and evaluate their expectations in order to have control which constants to be included in the model are estimable. When by the rules of 5.62 and 5.63, the number of fitted constants is made equal to N, apply the ranking method to the model containing the N constants.

6. Numerical Example

The non-orthogonal analysis of variance based upon a model constructed by the proposed procedure of fitting constants will numerically be demonstrated using the layout of Example 6. The data is artificial and was generated by the following model in which all effects involving factor $\mathscr{C}$ are absent:

$$\begin{aligned} y_{\alpha\beta\gamma} &= Y_{\alpha\beta\gamma} + e_{\alpha\beta\gamma} \\ &= \mu + a_{\alpha} + b_{\beta} + ab_{\alpha\beta} + e_{\alpha\beta\gamma} \quad \text{with} \quad e_{\alpha\beta\gamma} \sim NID\,(0, 1)\,. \end{aligned}$$

Because of the dummy factor $\mathscr{C}$, the analysis is expected to yield a significant model containing the effects $\mathscr{A}$, $\mathscr{B}$ and $\mathscr{A}\mathscr{B}$ only. The numerical values of the model constants for the construction of the data were chosen as follows: $\mu = 13$, $a_1 = 4$, $a_2 = 11$, $b_1 = -3$, $b_2 = 8$, $ab_{11} = 5$, $ab_{21} = -19$, $ab_{22} = 3$.

Using these values, the expected cell means $Y_{\alpha\beta\gamma}$ and the $n = 18$ actual 'observations' $y_{\alpha\beta\gamma}$ were constructed as shown in Fig. 7. For example, $Y_{111} = 13 + 4 - 3 + 5 = 19$. In 5 cells there are two repeated observations $Y_{\circ\beta\gamma}$ each such that a within-cell estimate of $\sigma^2 = 1$ can be obtained based on 5 degrees of freedom.

			$\mathscr{B}$		
		$\mathscr{C}$	$\beta = 1$	$\beta = 2$	$\beta = 3$
	$\alpha = 1$	$\gamma = 1$	$(Y_{111} = 19)$ $y_{1111} = 19.8$		$(Y_{131} = 17)$ $y_{1311} = 15.8$
		$\gamma = 2$	$(Y_{112} = 19)$ $y_{1121} = 18.2$ $y_{1122} = 20.7$		$(Y_{132} = 17)$ $y_{1321} = 15.7$ $y_{1322} = 16.1$
$\mathscr{A}$	$\alpha = 2$	$\gamma = 1$	$(Y_{211} = 2)$ $y_{2111} = 2.1$	$(Y_{221} = 35)$ $y_{2211} = 34.7$ $y_{2212} = 35.3$	
		$\gamma = 2$		$(Y_{222} = 35)$ $y_{2221} = 35.6$	$(Y_{232} = 24)$ $y_{2321} = 24.0$
	$\alpha = 3$	$\gamma = 1$	$(Y_{311} = 10)$ $y_{3111} = 10.4$ $y_{3112} = 10.8$		$(Y_{331} = 13)$ $y_{3311} = 13.0$
		$\gamma = 2$	$(Y_{312} = 10)$ $y_{3121} = 12.8$	$(Y_{322} = 21)$ $y_{3221} = 21.6$ $y_{3222} = 20.3$	$(Y_{332} = 13)$ $y_{3321} = 13.4$

Fig. 7
Data (layout) of Example 6) for numerical example.

Using the three models I, II and III constructed in the discussion of Example 6, the three ranking processes yield the results shown in Fig. 8. The calculations were performed using the program NOVACOM (Abt [2]).

Step No.	Effect	Fitted Constants Representing Effect	DF	$SSR_{N'}$ (Regression Sum of Squares)	$I(X)$
Model I ($bc_{21} = 0$)					
1	$\mathscr{ABC}$	abc_{111}	1	1383.039	0.385487
2	$\mathscr{BC}$	bc_{11}	1	1382.261	0.595450
3	$\mathscr{AC}$	ac_{11}, ac_{21}	2	1382.046	0.611865
4	$\mathscr{C}$	c_1	1	1380.544	0.571856
5	$\mathscr{AB}$	$ab_{11}, ab_{21}, ab_{22}$	3	1379.404	0.000202
6	$\mathscr{A}$	a_1, a_2	2	1014.176	0.000107
7	$\mathscr{B}$	b_1, b_2	2	802.219	0.000019
Model II ($ac_{21} = 0$)					
1	$\mathscr{ABC}$	abc_{111}	1	1383.039	0.385487
2	$\mathscr{BC}$	bc_{11}, bc_{21}	2	1382.261	0.700388
3	$\mathscr{AC}$	ac_{11}	1	1381.755	0.611865
4	$\mathscr{C}$	c_1	1	1380.544	0.571856
5	$\mathscr{AB}$	$ab_{11}, ab_{21}, ab_{22}$	3	1379.404	0.000202
6	$\mathscr{A}$	a_1, a_2	2	1014.176	0.000107
7	$\mathscr{B}$	b_1, b_2	2	802.219	0.000019
Model III ($ab_{21} = 0$)					
1	$\mathscr{ABC}$	abc_{111}	1	1383.039	0.385487
2	$\mathscr{AB}$	ab_{11}, ab_{22}	2	1382.261	0.025990
3	$\mathscr{AC}$	ac_{11}, ac_{21}	2	1363.350	0.000142
4	$\mathscr{BC}$	bc_{11}, bc_{21}	2	1095.730	0.000159
5	$\mathscr{C}$	c_1	1	1020.571	0.000202
6	$\mathscr{A}$	a_1, a_2	2	1014.176	0.000107
7	$\mathscr{B}$	b_1, b_2	2	802.219	0.000019

Fig. 8
Results of ranking processes in numerical example.

For the meaning of the results of the three ranking processes the reader is referred to Section 2 and to the discussion of Example 6 in Section 4. For instance, at Step 2 of the ranking process based on any of the three models, the three effects $\mathscr{AB}$, $\mathscr{AC}$, and $\mathscr{BC}$ are admissible for testing after the effect $\mathscr{ABC}$ (represented by abc_{111}) was ranked. For model III the $\mathscr{AB}$-interaction (represented by ab_{11} and ab_{22}) yields the largest $I(X)$-value among those of the three admissible effects: $\operatorname{Max} I(X) = 0.025990$.

This value of $I(X)$ corresponds to the F_c-value (5) with $n = 18$, $N = 12$, $N' = 9$, $SSR_N = 1383.039$, $SSR_{N'} = 1363.350$ and $\hat{\sigma}^2 = 0.862$:

$$F_c = \frac{1383.039 - 1363.350}{12 - 9} \Big/ 0.862 = 7.614$$

with 3 and 5 degrees of freedom. Under restricted admissibility, therefore, $\mathscr{AB}$ is ranked as the second-to-least-important effect. As can be seen when using Modell III, if the 'significant model' were to be determined at the level $\alpha = 0.05$, all effects except $\mathscr{ABC}$ would be considered 'significant' including those involving the dummy factor $\mathscr{C}$. Models I and II, however, lead to identical 'significant models' containing only the effects $\mathscr{A}$, $\mathscr{B}$ and $\mathscr{AB}$. Following the proposed procedure of choosing the most economical model when confounding is present, the analyst would come to the conclusion that a 7-degrees-of-freedom-model with $\mathscr{A}$, $\mathscr{B}$ and $\mathscr{AB}$ is sufficient to explain the differences between the 13 occupied cells.

Naturally, the analyst would not have known that this model was exactly the one used to construct the data. It appears interesting how close to the true values the $1 + 7 = 8$ estimates of the model constants resulted in step 5 for both Models I and II:

$$\hat{\mu} = 13.2,$$

$$\hat{a}_1 = 2.7, \quad \hat{a}_2 = 10.8, \quad \hat{b}_1 = -1.9, \quad \hat{b}_2 = 7.8,$$

$$\widehat{ab}_{11} = 5.6, \; \widehat{ab}_{21} = -20, \; \widehat{ab}_{22} = 3.4.$$

References

[1] ABT, K. (1967): On the Identification of the Significant Independent Variables in Linear Models. Metrika *12*, 1–15, 81–96.

[2] ABT, K. (1968): The Method and Use of NOVACOM, a Program for Non-Orthogonal Analysis of Variance and Covariance. Report No. 2108, U.S. Naval Weapons Laboratory, Dahlgren, Virginia.

[3] BOCK, R. D. (1963): Programming Univariate and Multivariate Analysis of Variance. Technometrics *5*, 95–117.

[4] DYKSTRA, O., Jr. (1966): The Orthogonalization of Undesigned Experiments. Technometrics *8*, 279–290.

[5] ELSTON, R. C., and BUSH, N. (1964): The Hypotheses That Can Be Tested When There Are Interactions in an Analysis of Variance Model. Biometrics *20*, 681–698.

[6] FEDERER, W. T. (1957): Variance and Covariance Analyses for Unbalanced Classifications. Biometrics *13*, 333–362.

[7] GORMAN, J. W., and TOMAN, R. J. (1966): Selection of Variables for Fitting Equations to Data. Technometrics *8*, 27–51.

[8] GRAYBILL, F. A. (1961): An Introduction to Linear Statistical Models, Volume I. McGraw-Hill Book Co., Inc., New York.

[9] SCHEFFÉ, H. (1959): The Analysis of Variance. John Wiley & Sons, Inc., New York.

[10] SELVIN, H., C. and STUART, A. (1966): Data-Dredging Procedures in Survey Analysis. The American Statistician *20*, 20–23.

[11] YATES, F. (1934): The Analysis of Multiple Classifications with Unequal Numbers in the Different Classes. Journal of the American Statistical Association *29*, 51–66.

Author's address:
Prof. Dr. Klaus Abt, Abteilung für Biomathematik, Fachbereich Humanmedizin, Universität Frankfurt a. M.

H. Linhart

H. Linhart

The Random Analysis of Variance Model and the Wrong Test

Summary

In random analysis of variance one uses the statistic MSA/MSAB to test the hypothesis that 'factor A has no effect'. Here the *wrong* test, using MSA/MSE as statistic, is compared with the usual test. The two tests do, of course, test two different hypotheses. An analysis of the employed terminology reveals that the *wrong* hypothesis is not 'worse' than the usual hypothesis. It is analogous to the corresponding hypothesis in *fixed* analysis of variance. Because of that, and by power considerations, it is argued that the *wrong* test should be preferable in most applications.

1. **Introduction**

For some time I have had the feeling that one should scrutinise more carefully the logic of the generally accepted tests of significance in random analysis of variance models. It seemed intuitively difficult to understand why one should apply different tests of the hypothesis 'that a factor has no effect', depending on whether the levels have been determined by some non-random procedure, or whether they resulted from random sampling from a population of levels. In particular the analogy to linear regression comes to one's mind, and one wonders whether one cannot also in random, crossed, analysis of variance apply a conditional test, which would in fact be the usual test for fixed models. I discussed this with Professor A. Linder in 1967, while he was Visiting Professor in our University. He encouraged me to pursue this question and mentioned a paper by Yates [4] which seemed to express similar ideas.

The detailed discussion which follows leads to the recommendation to use the same test of the hypothesis 'that a factor has no effect', regardless of whether a fixed or a random model is assumed. If the recommendation is followed it is not necessary—for purposes of this test—to chose between the two models. The recommended test is the one which is generally accepted for the fixed model.

It must be understood that this is essentially a recommendation to test a certain hypothesis rather than another one. I think that this recommendation will be sound in the majority of practical applications but it cannot be stated categorically that it will be sound in *all* applications. (The same reservation holds, however, also for the generally accepted test if the fixed model is assumed.)

At first the random and the fixed analysis of variance models are stated and the usual tests, using the statistics MSA/MSAB (random test) and MSA/MSE (fixed test) respectively, are discussed. The 'wrong' test, assuming that the random model holds and using MSA/MSE as statistic, is also considered. The random and the fixed test seem to test the 'effect of A', whereas the wrong test seems not to test the 'effect of A'.

It is then argued that this impression is misleading and mainly a question of terminology. The random and the wrong test do definitely test different hypotheses, but the 'wrong hypothesis' is by no means inferior to the 'random hypothesis' it is also completely analogous to the 'fixed hypothesis', the usual hypothesis tested if a fixed model is assumed.

The power functions of the random and the wrong test are then studied. They are power functions of tests of different hypotheses and are therefore not comparable. But the results indicate that, in a certain sense, the wrong test is a more powerful test of the wrong hypothesis than the random test of the random hypothesis.

2. Two Models

At the outset we give for later reference the random (1), and the fixed (2, 3), crossed, analysis of variance model which will be used in this paper.

The random model is

$$y_{ijk} = \mu + a_i + b_j + (a\,b)_{ij} + e_{ijk}\,,$$

$$i = 1, 2, \ldots, I; \quad j = 1, 2, \ldots, J; \quad k = 1, 2, \ldots, K; \tag{1}$$

where a_i, b_j, $(a\,b)_{ij}$, e_{ijk} are mutually independently and normally distributed with zero mean and with variances σ_a^2, σ_b^2, σ_{ab}^2 and σ_e^2 respectively. This model leads to the analysis of variance and the distributions given in Table 1. The sums of squares for the mean, A, B, $A\,B$, E, are mutually independently distributed.

Table 1
Analysis of Variance.

Source	d. f.	SS	Distribution of SS
Mean	1	$\Sigma \bar{y}_{...}^2$	$(J\,K\,\sigma_a^2 + I\,K\,\sigma_b^2 + K\,\sigma_{ab}^2 + \sigma_e^2)\,\chi^2(1)$
A	$I-1$	$\Sigma(\bar{y}_{i..} - \bar{y}_{...})^2$	$(J\,K\,\sigma_a^2 + K\,\sigma_{ab}^2 + \sigma_e^2)\,\chi^2(I-1)$
B	$J-1$	$\Sigma(\bar{y}_{.j.} - \bar{y}_{...})^2$	$(I\,K\,\sigma_b^2 + K\,\sigma_{ab}^2 + \sigma_e^2)\,\chi^2(J-1)$
$A\,B$	$(I-1)\,(J-1)$	$\Sigma(\bar{y}_{ij.} - \bar{y}_{i..} - \bar{y}_{.j.} + \bar{y}_{...})^2$	$(K\,\sigma_{ab}^2 + \sigma_e^2)\,\chi^2((I-1)\,(J-1))$
E	$I\,J(K-1)$	$\Sigma(y_{ijk} - \bar{y}_{ij})^2$	$\sigma_e^2\,\chi^2(I\,J(K-1))$
Total		$\Sigma\,y_{ijk}^2$	

The sums are over all three indices: i, j and k.

We write $\chi^2(\nu;\lambda)$ for the noncentral chi-squared distribution with ν degrees of freedom and with noncentrality parameter λ, also $\chi^2(\nu;0) \equiv \chi^2(\nu)$.

The fixed model is given by

$$y_{ijk} = \mu + \alpha_i + \beta_j + (\alpha\,\beta)_{ij} + e_{ijk} \tag{2}$$

with the constraints

$$\sum_i \alpha_i = \sum_j \beta_j = \sum_j (\alpha\,\beta)_{ij} = \sum_i (\alpha\,\beta)_{ij} = 0 \tag{3}$$

(for all i and all j), where the e_{ij} are $NID(0, \sigma_e^2)$.

3. Three Possible Tests

If the random model (1) is assumed, the usual test of the hypothesis $\sigma_a^2 = 0$ is provided by the test statistic MSA/MSAB which has, under the hypothesis, an F-distribution with $I-1$ and $(I-1)\,(J-1)$ degrees of freedom.

By analogy to regression analysis one could try to test the hypothesis $a_i = 0$ $(i = 1, 2, \ldots, I)$ conditionally, given b_j and $(a\,b)_{ij}$ $(i = 1, 2, \ldots, I;$ $j = 1, 2, \ldots, J)$. (It is, of course, loose terminology to speak about the 'hypothesis' that $a_i = 0$, $i = 1, 2, \ldots, I$. We use it since comparison with the fixed model is facilitated in this way. The hypothesis $\sigma_a^2 = 0$ and the 'hypothesis' $a_i = 0$, $i = 1, 2, \ldots, I$, are equivalent in the following sense: If $\sigma_a^2 = 0$, then $P(a_i = 0, i = 1, 2, \ldots, I) = 1$. If $a_i = 0$, $i = 1, 2, \ldots, I$, then the hypothesis that $\sigma_a^2 \neq 0$ can be rejected at any desired significance level since $P(a_i = 0, i = 1, 2, \ldots, I) = 0$ if $\sigma_a^2 \neq 0$. Similar remarks apply later in this paper). One might be of the opinion that this is done if one proceeds as if (2, 3) were true and one wanted to test $\alpha_i = 0$, $i = 1, 2, \ldots, I$, that is: if one uses MSA/MSE as test statistic. This is, however, not so as the following argument shows.

If a_i, b_j, $(a\,b)_{ij}$, $(i = 1, 2, \ldots, I;\ j = 1, 2, \ldots, J)$ are fixed (1) is not equivalent to (2, 3), since the constraints (3) are not satisfied. The only way to reduce the conditional model to the fixed model (2, 3) seems to be the following.

Write

$$y_{ijk} = \big(\mu + \bar{a}. + \bar{b}. + \overline{(a\,b)}..\big) + \big(a_i + \overline{(a\,b)}_{i.} - \bar{a}. - \overline{(a\,b)}..\big)$$

$$+\big(b_j + \overline{(a\,b)}_{.j} - \bar{b}. - \overline{(a\,b)}..\big) + \big((a\,b)_{ij} - \overline{(a\,b)}_{i.} - \overline{(a\,b)}_{.j} + \overline{(a\,b)}..\big)$$

$$+ e_{ijk} \tag{4}$$

and set the expressions in brackets identical to μ, α_i, β_j, $(\alpha\,\beta)_{ij}$, respectively, then the constraints (3) are satisfied and model (2, 3) holds.

The test which uses MSA/MSE as statistic is, therefore, considered as a conditional test, given b_j, $(a\,b)_{ij}$, a test of the hypothesis that

$$a_i + \overline{(a\,b)}_{i.} - \bar{a}. - \overline{(a\,b)}.. = 0 \quad \text{for} \quad = 1, 2, \ldots, I\,.$$

The models, hypotheses, statistics and names of the three possible tests are displayed in Table 2.

Table 2
Possible Tests.

Assumed	Hypothesis	Statistic	Name
(1)	$a_i = 0$	MSA/MSAB	Random
(1)	$a_i + \overline{(a\,b)}_{i.} - \bar{a}_{.} - \overline{(a\,b)}_{..} = 0$	MSA/MSE	Wrong
(2, 3)	$\alpha_i = 0$	MSA/MSE	Fixed

In all cases $i = 1, 2, \ldots, I$.

To have a convenient terminology we shall speak about the random test (and the random hypothesis) the wrong test (and the wrong hypothesis) and the fixed test (and the fixed hypothesis).

The random and the fixed test seem to test the 'effect of A', whereas the wrong test seems not to test the 'effect of A'. In the next section we shall show that this impression is misleading and mainly a question of terminology. The random and the wrong test do definitely test different hypotheses, but we shall show that the wrong hypothesis is in no way 'inferior' to the random hypothesis. (Of course it is not possible to say that one hypothesis is better or worse than the other. But a superficial look at the wrong hypothesis creates the impression that it is not likely to be of much interest in practice.)

4. The Fixed Model

If one approaches an experimental situation by writing down (2) one would, in fact, like to estimate, and test, α_i, β_i and $(\alpha\,\beta)_{ij}$ from the data. One realises that this is not possible, and the difficulty is solved by postulating the constraints (3). The assumptions (3) seem, however, to be entirely *ad hoc*, they are not justified by the experimental situation. Estimation and testing of α_i, β_j and $(\alpha\beta)_{ij}$ under the constraints (3) seems to be a compromise. One salvages something where not everything is attainable. I seem to say now that our classical approach is making unwarranted assumptions. I do not, in fact, since the classical approach, as e.g. outlined in Kempthorne [2], is not to write down model (2). The approach is via the one-way analysis of variance model

$$y_{ijk} = \theta_{ij} + e_{ijk} . \tag{5}$$

There is no difficulty here in estimating the treatment effects θ_{ij}. One defines then

$$\mu = \bar{\theta}_{..}, \quad \alpha_i = \bar{\theta}_{i.} - \bar{\theta}_{..}, \quad \beta_j = \bar{\theta}_{.j} - \bar{\theta}_{..},$$

$$(\alpha\beta)_{ij} = \theta_{ij} - \bar{\theta}_{i.} - \bar{\theta}_{.j} + \bar{\theta}_{..} . \tag{6}$$

With these definitions model (2) with the constraints (3) holds.

The classical approach is therefore eminently reasonable. Is it then at all reasonable to approach an experimental situation by writing down (2) without the constraints (3)? I think it is, only one cannot, *from the usual $I\,J\,K$ observations in a factorial experiment*, estimate the 'effects' and 'interactions'.

Let us consider then, what one would call the effect of A (at two levels, 1 and 2, say) the effect of B (at two levels, say) and the interactions of A and B (at the 4 possible combinations of levels) if one would approach the situation naively, having never heard about factorial experiments and their terminology. We shall denote these *naive* effects and interactions by α_i^*, β_j^*, $(\alpha\beta)_{ij}^*$, in contrast to the *traditional* effects and interactions $\alpha_i, \beta_j, (\alpha\beta)_{ij}$.

Assume that one is dealing with a chemical process and is measuring a variable y, and that A and B are two different additives (two different amounts each, corresponding to the two levels). If the expectation of y without any additives is μ^* and the expectation of y with A only, at level i, is $\mu^* + \alpha_i^*$, then one feels justified to call μ^* and α_i^* the naive overall mean and the naive effect of A at level i. Similarly, if the expectation with additive B only, at level j, is $\mu^* + \beta_j^*$ one would naively call β_j^* the effect of B at level j. If the expectation of y in the case where both additives are used (A at level i, B at level j) is $\mu^* + \alpha_i^* + \beta_j^* + (\alpha\beta)_{ij}^*$, one could call $(\alpha\beta)_{ij}^*$ the naive interaction of A at level i with B at level j. The observations in a 2^2 factorial experiment comprise the $4\,K$ observations belonging to K replications of the experiment where both additives were used, at the 4 possible combinations of levels. With such observations it is not possible to estimate α_i^*, β_j^* and $(\alpha\beta)_{ij}^*$. (One could estimate these constants with additional observations: without additives, with A only, at levels 1 and 2, with B only, at levels 1 and 2.)

If one sets $\theta_{ij} = \mu^* + \alpha_i^* + \beta_j^* + (\alpha\beta)_{ij}^*$ then one gets from (5)

$$
\begin{aligned}
\mu &= \mu^* + \bar{\alpha}.^* + \bar{\beta}.^* + \overline{(\alpha\beta)}..^*, \\
\alpha_i &= \alpha_i^* - \bar{\alpha}.^* + \overline{(\alpha\beta)}_{i.}^* - \overline{(\alpha\beta)}..^*, \\
\beta_j &= \beta_j^* - \bar{\beta}.^* + \overline{(\alpha\beta)}_{.j}^* - \overline{(\alpha\beta)}..^*, \\
(\alpha\beta)_{ij} &= (\alpha\beta)_{ij}^* - \overline{(\alpha\beta)}_{i.}^* - \overline{(\alpha\beta)}_{.j}^* + \overline{(\alpha\beta)}..^*.
\end{aligned} \tag{7}
$$

It is now clear that the hypothesis $\alpha_i = 0$, $i = 1, 2, \ldots, I$, is not the hypothesis 'that A has no naive effect'. The fixed hypothesis is thus, in this sense, not superior to the wrong hypothesis in the analysis of the random model.

5. **The Random Model**

The random model also needs further scrutiny. The basic random variables determining the distributions of (naive and traditional) random effects and interactions are the random levels of factors A and B, we shall denote them by x and y. Effects and interactions are functions of x and y and we shall, for any function $g(x, y)$ of x and y, use the notation $g(x_i, y_j) \equiv g_{ij}$. The index i will always be associated with variable x and the index j with variable y. If i (or j) does not appear as index this indicates that the function does not depend on x (or y). (E.g. h_j indicates that $h(x, y) \equiv h(y)$, $h(y_j) \equiv h_j$). Denote by μ^* the expected yield in the absence of factors A and B. (The expectations in this paragraph are always taken over replications.) If now factor A is added, at a randomly chosen level x_i, denote the expectation by $\mu^* + a_i^*$ and call a_i^* the naive effect of factor A at level i. Similarly, if $\mu + b_j^*$ is the expected yield if factor B only is added, at a randomly chosen level y_j, call b_j^* the naive effect of factor B at level j. If factor A, at a randomly chosen level x_i, and factor B, at a randomly chosen level y_j, are added let the expected yield be t_{ij}. Then one must obviously define the naive, random, interaction $(a\,b)_{ij}^*$ by

$$t_{ij} = \mu^* + a_i^* + b_j^* + (a\,b)_{ij}^*. \tag{8}$$

It should usually be acceptable to assume that the levels of A and B are independently chosen, i.e. that x and y are independently distributed. It follows then that $a^*(x)$ and $b^*(y)$ are independently distributed. It is difficult to make any further assumptions about the distributions of a^* and b^* and in particular about the distribution of $(a\,b)^*$. Fortunately this is not necessary for the arguments which follow. Scheffé [3], in a very careful analysis, came to the conclusion that model (1) with its distributional assumptions, in particular the independence of $a(x)$, $b(y)$, $(a\,b)\,(x, y)$, is reasonable if one is prepared to assume normality.

He, in fact, defines

$$\begin{aligned}
a_i &= \underset{y}{E}\big(t(x_i, y)\big) - E\big(t(x, y)\big), \\
b_j &= \underset{x}{E}\big(t(x, y_j)\big) - E\big(t(x, y)\big), \\
(a\,b)_{ij} &= t(x_i, y_j) - \underset{y}{E}\big(t(x_i, y)\big) - \underset{x}{E}\big(t(x, y_j)\big) + E\big(t(x, y)\big). \qquad (9)
\end{aligned}$$

One gets then from (8)

$$\begin{aligned}
a_i &= a_i^* - E\big(a^*(x)\big) + \underset{y}{E}\big((a\,b)^*\,(x_i, y)\big) - E\big((a\,b)^*\,(x, y)\big), \\
b_j &= b_j^* - E\big(b^*(y)\big) + \underset{x}{E}\big((a\,b)^*\,(x, y_j)\big) - E\big((a\,b)^*\,(x, y)\big), \\
(a\,b)_{ij} &= (a\,b)_{ij}^* - \underset{y}{E}\big((a\,b)^*\,(x_i, y)\big) - \underset{x}{E}\big((a\,b)^*\,(x, y_j)\big) \qquad (10) \\
&\quad + E\big((a\,b)^*\,(x, y)\big).
\end{aligned}$$

Therefore the hypothesis that $a_i = 0$, $i = 1, 2, \ldots, I$, is equivalent to the hypothesis that

$$a_i^* - E\big(a^*(x)\big) + \underset{y}{E}\big((a\,b)^*\,(x_i, y)\big) - E\big((a\,b)^*\,(x\;y)\big) = 0$$

for $i = 1, 2, \ldots, I$.

The wrong hypothesis is that $a_i + \overline{(a\,b)}_{i.} - \bar{a}_. - \overline{(\bar{a}\,\bar{b})}_{..} = 0$, $i = 1, 2, \ldots, I$. This, rewritten in terms of naive effects, becomes

$$a_i^* + \overline{(a\,b)}_{i.}^* - \bar{a}_.^* - \overline{(a\,b)}_{..}^* = 0,\ i = 1, 2, \ldots, I.$$

We state once more the three hypotheses:

$$\begin{aligned}
&\text{Random: } a_i^* - E\big(a^*(x)\big) + \underset{y}{E}\big((a\,b)^*\,(x_i, y)\big) - E\big((a\,b)^*\,(x, y)\big) = 0, \\
&\text{Wrong: } \quad a_i^* - \bar{a}_.^* + \overline{(a\,b)}_{i.}^* - \overline{(a\,b)}_{..}^* = 0, \qquad (11) \\
&\text{Fixed: } \quad \alpha_i^* - \bar{\alpha}_.^* + \overline{(\alpha\,\beta)}_{i.}^* - \overline{(\alpha\,\beta)}_{..}^* = 0,
\end{aligned}$$

(in all cases for $i = 1, 2, \ldots, I$).

None of the three tests does, in fact, test the hypothesis that the naive effect of A is the same at all levels. There is, therefore, very little to choose between the random and the wrong test. None is entirely satisfactory and I personally would be inclined to choose the wrong test because its hypothesis is completely analogous to the hypothesis in the fixed model.

It seems natural to investigate the power of the two tests in the hope that this might help to judge the issue. We do this in the next section.

6. Power Functions

The random test

From Table 1 follows that $F_R = \mathrm{MSA}/\mathrm{MSAB}$ is distributed as $\big((J\,K\,\sigma_a^2 + K\,\sigma_{ab}^2 + \sigma_e^2)/(K\,\sigma_{ab}^2 + \sigma_e^2)\big)\,F\big(I-1,\,(I-1)\,(J-1)\big)$. Now set $\theta_R = \sigma_a^2$ and $\lambda_R = 1/(\sigma_{ab}^2/J + \sigma_e^2/J\,K)$, write $F^\alpha(\gamma_1, \gamma_2)$ for the 100 α per cent point of
$F(\gamma_1, \gamma_2)$ and $P(\theta, \lambda, \gamma_1, \gamma_2, \alpha) = \text{Prob.}\big(F(\gamma_1, \gamma_2) > F^\alpha(\gamma_1, \gamma_2)/(1 + \lambda\,\theta)\big)$.
Then the power function of the random test is

$$P\big(\theta_R, \lambda_R, I-1, (I-1)\,(J-1), \alpha\big)\,. \tag{12}$$

The wrong test

It is not difficult to establish that the power function of the wrong test is

$$P(\theta_W, \lambda_W, I-1, I\,J(K-1), \alpha)\,,$$

where

$$\theta_W = \frac{\sigma_a^2 + \sigma_{ab}^2}{J} \quad \text{and} \quad \lambda_W = \frac{J\,K}{\sigma_e^2}\,. \tag{13}$$

Comparison

One can now prove the following:

a) For every fixed set of values $\sigma_a^2, \sigma_{ab}^2, \sigma_e^2$ the wrong test is more powerful than the random test.

b) If $\sigma_{ab}^2 = 0$, the wrong and the random hypothesis coincide, and because of a) the wrong test is more powerful. (This follows since $P(\theta, \lambda, \gamma_1, \gamma_2, \alpha)$ is monotonically increasing with increasing θ and λ and because $\theta_R \le \theta_W$ and $\lambda_R \le \lambda_W$, where in both cases the equality sign holds if and only if $\sigma_{ab}^2 = 0$).

7. Conclusion

As stated before, we are in favour of using the wrong test to test the wrong hypothesis. It was established in Sections 3 and 4 that the wrong hypothesis is entirely analogous (in the naive sense) to the fixed hpyothesis.

Also, a) above can surely be interpreted in the sense that the wrong test is a more powerful test of the wrong hypothesis than the random test of the random hypothesis.

And point b) above shows that for $\sigma_{ab}^2 = 0$ the wrong test is clearly preferable.

Hence: if one has no clear preference for one or the other hypothesis, the power considerations should turn the balance in favour of the wrong test.

8. Acknowledgement

After this paper was written, A. A. Rayner pointed out to me that Anderson ([1], Section 5, p. 49) has introduced a new model for factorial experiments which—apart from minor differences—amounts to using 'naive' and not 'traditional' effects and interactions. This was done in an entirely different context.

I would like to thank A. A. Rayner for this reference and for other remarks which lead to improvements in the paper.

References

[1] ANDERSON, R. L. (1960): Some remarks on the design and analysis of factorial experiments. In: I. Olkin et al., eds. Contributions to Probability and Statistics. Stanford Univ. Press, Stanford, 35–36.

[2] KEMPTHORNE, O. (1952): The design and analysis of experiments. J. Wiley, New York.

[3] SCHEFFÉ, H. (1956): Alternative models for the analysis of variance. Ann. Math. Statist. *27*, 251–271.

[4] YATES, F. (1967): A fresh look at the basic principles of the design and analysis of experiments. Proc. Fifth Berkeley Symposium on Mathematical Statistics and Probability *4*, 777–790.

Author's address:
H. Linhart, University of Natal, King George V Avenue, Durban, South Africa.

B. M. Bennett

B. M. Bennett

On an Approximate Test for Homogeneity of Coefficients of Variation

Summary

The coefficient of variation (c. v.) $\zeta = \sigma/\xi$, or ratio of standard deviation to mean ($\xi > 0$), is a useful parameter and measure of variability in certain situations since it is expressed in absolute units. Thus in comparing results on repeated clinical data when k (≥ 2) different measurement techniques are utilized in analyzing the same or paired specimens, it is of interest to compare the corresponding sample c. v.'s $z_1, \ldots, z_k$, respectively for the k techniques. For this situation the following note presents a test which is based on McKay's approximation [4] to the distribution of the sample c. v. for the normal distribution, and utilizes Pitman's method [5] for the hypothesis of equality of scale parameters of Gamma variates.

1. Introduction

Suppose $\{x_i\}$ ($i = 1, \ldots, n$) represent a sample of n independent observations from some continuous distribution f with mean ξ (> 0) and variance σ^2. The c. v. $\zeta = \sigma/\xi$ is a useful measure of variability since it is expressed in absolute units.

If f happens to be the normal distribution $N(\xi, \sigma)$, then the sample estimate of ζ, $z = s/\bar{x}$, defined in terms of $n\,\bar{x} = \sum_i x_i$, $(n-1)\,s^2 = \Sigma(x_i - \bar{x})^2$, is related to the non-central t-distribution since $t = \sqrt{n}/z$. Tang [6] gave a series expansion for the distribution of z. Various authors (Cf. references in Koopmans, Owen and Rosenblatt [3]) have presented results, sometimes incorrect, for approximations to the distribution of z.

Recently Iglewicz et al. [1, 2] have shown that McKay's approximation [4] of the transformed variable

$$\frac{y}{\phi} = \frac{\left(\dfrac{n\,z^2}{1+z^2}\right)}{\left(\dfrac{\zeta^2}{1+\zeta^2}\right)} \tag{1}$$

by a corresponding χ^2 distribution with $(n-1)$ D. F. is in fact quite an accurate one. This approximation assumes that the probability of a negative z value (corresponding to negative $\bar{x}$) is negligible.

2. Approximate Test for Homogeneity of C.V.'s in Repeated Samples

Let then $\{x_{ij}\}$ $(i = 1, \ldots, k; j = 1, \ldots, n_i)$ represent $n = \Sigma n_i$ independent observations from k normal populations $N(\xi_i, \sigma_i^2)$ with c. v.'s $\zeta_i = \sigma_i/\xi_i$, where $\xi_i > 0$. It is required to test the hypothesis H_0 of homogeneity, or

$$H_0 : \zeta_1 = \cdots = \zeta_k = \zeta \quad \text{(unspec.)} . \tag{2}$$

We set

$$\frac{y_i}{\phi_i} = \frac{\dfrac{n_i z_i^2}{1 + z_i^2}}{\dfrac{\zeta_i^2}{1 + \zeta_i^2}} ,$$

defined in terms of

$$z_i = \frac{s_i}{\bar{x}_{i.}} , \quad n_i x_{i.} = \sum_j x_{ij} , \quad (n_i - 1) s_i^2 = \sum_j (x_{ij} - \bar{x}_{i.})^2 .$$

Testing the hypothesis H_0 is equivalent to that of the equality of the scale parameters, $\phi_1 = \cdots = \phi_k$ $(= \phi)$ for the k independent Gamma variates $y_1, \ldots, y_k$ (Pitman [5].)

The appropriate (Pitman) statistic for H_0 is then

$$2Z = (n-k) \log\left\{\frac{\sum_i y_i}{n-k}\right\} - \sum (n_i - 1) \log\left\{\frac{y_i}{n_i - 1}\right\}$$

$$= (n-k) \log\left\{\frac{\sum \dfrac{n_i z_i^2}{1+z_i^2}}{n-k}\right\} - \sum (n_i - 1) \log\left\{\frac{\dfrac{n_i z_i^2}{1+z_i^2}}{n_i - 1}\right\} \tag{3}$$

which is approximately distributed as χ^2 with $(k-1)$ D. F.

References

[1] IGLEWICZ, B., MYERS, R. H. and HOWE, R. B. (1968): On the percentage points of the sample coefficient of variation. Biometrika *56*, 580–581.

[2] IGLEWICZ, B. and MYERS, R. H. (1970): Comparisons of approximations to the percentage points of the sample coefficient of variation. Technometrics *12*, 166–169.

[3] KOOPMANS, L. H., OWENS, D. B. and ROSENBLATT, J. I. (1964): Confidence intervals for the coefficient of variation for the normal and log-normal distributions. Biometrika *51*, 25–32.

[4] MCKAY, A. T. (1932): Distribution of the coefficient of variation and the extended t distribution. J. R. Statist. Sec. *95*, 695–698.

[5] PITMAN, E. J. G. (1939): Tests of hypotheses concerning location and scale parameters. Biometrika *31*, 200–215.

[6] TANG, P. C. (1938): The power function of the analysis of variance tests with tables and illustrations of their use. Stat. Res. Mem. *2*, 126–149.

Key Words: Coefficients of variation; in independent samples;
McKay's approximation; Pitman's test for scale parameters.

Author's address:
B. M. Bennett, University of Hawaii, School of Public Health, Honolulu.

W. Berchtold

W. Berchtold

Analyse eines Cross-Over-Versuches mit Anteilziffern

1. Einleitung

Die Analyse von Cross-over-Versuchen mit zwei Verfahren und normal verteilten Residuen ist gut bekannt und in vielen Lehrbüchern über Versuchsplanung – zum Beispiel bei Cochran and Cox [2] – beschrieben. Dem Versuch liegt ein lineares Modell zugrunde, das die Verfahren, die Versuchsperiode und die Blockeffekte (in einem klinischen Versuch zum Beispiel bilden die Patienten die Blöcke) berücksichtigt. Die Streuungszerlegung der Messwerte hat folgenden Aufbau:

Streuung	*Freiheitsgrad*
Zwischen Blöcken	$b-1$
Innerhalb Blöcken:	
Verfahren	1
Perioden	1
Rest	$b-2$
Insgesamt	$2 \cdot b - 1$

Im folgenden beschränken wir uns auf die Anteile «innerhalb der Blöcke».

Liegen anstelle gemessener Werte Anteilziffern π vor, so darf die Streuungszerlegung nicht mehr in der üblichen Art durchgeführt werden; die Voraussetzungen des linearen Modelles sind dann verletzt. Die Wirkung von Verfahren und Versuchsperiode ist nur in einem engen Bereich genähert additiv. Die Streuung ist nicht für alle Messwerte gleich; sie hängt von π ab und ist gegeben durch $\sigma^2 = \pi(1-\pi)/N$ mit N = Zahl der Beobachtungen.

Will man trotzdem die bewährte Methode der Streuungszerlegung anwenden, so müssen die Anteilziffern in geeignete Grössen z transformiert werden. Die Transformation ist so zu wählen, dass die Effekte additiv werden. Die Inhomogenität der Streuungen wird durch Gewichtsfaktoren berücksichtigt.

Üblich sind die Probit-, die Logit-, die komplementäre Loglog- und die Wurzeltranformation. Die Wahl der geeigneten Transformation ist ausführlich beschrieben bei Linder [3]. Die Streuungszerlegung mit Anteilziffern kann nachgelesen werden bei Berchtold und Linder [1]. Das dort dargelegte Verfahren muss für den Cross-over-Versuch modifiziert werden.

Dieser Arbeit liegt das folgende praktische Problem zugrunde: Zu prüfen ist die Wirkung eines Medikamentes gegen Kopfweh, Migräne oder epileptische Anfälle. Die Patienten (total b) werden zufällig in zwei Gruppen mit $b1$ bzw. $b2$ Patienten aufgeteilt. Die erste Gruppe erhält in der ersten Versuchsperiode ein Placebo und in der zweiten Periode ein Testpräparat; bei der zweiten Gruppe ist die Reihenfolge umgekehrt. Die beobachtete Zeitdauer sei N_{ij} Tage beim i-ten Patienten in der j-ten Periode. Die Zahl der Tage mit Anfällen wird mit x_{ij} bezeichnet. Der Quotient x_{ij}/N_{ij} ist ein Schätzwert für die Anfallswahrscheinlichkeit π_{ij}. Man vermutet, dass π_{ij} unter der Wirkung des Testpräparates kleiner ist als unter der Wirkung des Placebos. Gewöhnlich gelingt es nicht, bei allen Patienten über gleich viele Tage zu beobachten, so dass N_{ij} nicht für alle i und j konstant bleibt. Die Zahl der Anfallstage x_{ij} ist binomisch verteilt, die Streuung wird also abhängig von der Anfallswahrscheinlichkeit.

In Kapitel 2 werden die Likelihoodschätzungen für die Parameter des linearen Modelles und in Kapitel 3 Testgrössen zum Prüfen der Hypothesen $\alpha = 0$, $\gamma = 0$ und über die Zulässigkeit des Modelles angegeben. In Kapitel 4 wird das Verfahren in einem Zahlenbeispiel angewendet. Die Bezeichnungen in den Kapiteln 2 und 3 werden diesem Beispiel angepasst.

2. Likelihoodschätzungen

Den beiden Behandlungsreihenfolgen werden $b1$ bzw. $b2$ Patienten (Blöcke) zufällig zugeteilt. Im allgemeinen wird $b1$ von $b2$ verschieden sein; $b1 + b2 = b$. Die Beobachtungsdauer (in Tagen) wird mit N_{ij}, die Zahl der Tage mit Anfällen mit x_{ij} bezeichnet. Dabei gibt der Index i den Patienten und j, $j = 1$ oder 2, die Versuchsperiode an. Die Likelihoodfunktion ist ein Produkt von Binomialverteilungen; sie kann sofort hingeschrieben werden:

$$L = \prod_{i=1}^{b} \prod_{j=1}^{2} \binom{N_{ij}}{x_{ij}} \pi_{ij}^{x_{ij}} (1 - \pi_{ij})^{N_{ij}-x_{ij}} . \tag{2.1}$$

Im folgenden wird immer mit dem Logarithmus von L gerechnet:

$$\ln L = \sum_{i=1}^{b} \sum_{j=1}^{2} x_{ij} \ln \pi_{ij} + \sum_{i=1}^{b} \sum_{j=1}^{2} (N_{ij} - x_{ij}) \ln (1 - \pi_{ij}) + \text{konst.} \tag{2.2}$$

Dabei bedeuten

$$\pi_{ij} = f^{-1}(z_{ij}) \quad \text{oder} \quad z_{ij} = f(\pi_{ij}) . \tag{2.3}$$

Für z gelte das lineare Modell

$$E(z_{ij}) = \beta_i + \delta_{ij}^{(\alpha)} \alpha + \delta_i^{(\gamma)} \gamma \tag{2.4}$$

mit β_i = Patientenparameter, α = Behandlungsparameter, γ = Periodenparameter.

Die Grössen δ sind wie folgt definiert:

$$\delta_{ij}^{(\alpha)} = \begin{rcases} -1 \text{ für } j = 1 \\ +1 \text{ für } j = 2 \end{rcases} \text{ Patient } i \text{ aus der Gruppe Placebo} \to \text{Medikament,}$$

$$\begin{rcases} +1 \text{ für } j = 1 \\ -1 \text{ für } j = 2 \end{rcases} \text{ Patient } i \text{ aus der Gruppe Medikament} \to \text{Placebo,}$$

$$\delta_j^{(\gamma)} = (-1)^j, \text{ das heisst } = \begin{array}{l} -1 \text{ in der ersten Periode,} \\ +1 \text{ in der zweiten Periode.} \end{array}$$

Die Schätzungen für die Parameter β_i, α und γ sind Nullstellen des Systems der ersten Ableitungen von $\ln L$.

Es ist zweckmässig, die folgende Definition einzuführen:

$$d_{ij} = \left[\frac{x_{ij}}{\pi_{ij}} - \frac{N_{ij} - x_{ij}}{1 - \pi_{ij}}\right] \left(\frac{\partial \pi_{ij}}{\partial z_{ij}}\right)$$

$$= \left[\frac{x_{ij} - N_{ij} \cdot \pi_{ij}}{\pi_{ij}(1 - \pi_{ij})}\right] \left(\frac{\partial \pi_{ij}}{\partial z_{ij}}\right). \tag{2.5}$$

Damit können die ersten Ableitungen in die nachstehende knappe Form gebracht werden:

$$\frac{\partial \ln L}{\partial \beta_i} = d_{i1} + d_{i2} = 0\,, \quad i = 1, 2 \ldots b\,, \tag{2.6}$$

$$\frac{\partial \ln L}{\partial \alpha} = \sum_i \sum_j \delta_{ij}^{(\alpha)} d_{ij} = 0\,, \tag{2.7}$$

$$\frac{\partial \ln L}{\partial \gamma} = \sum_i (d_{i2} - d_{i1}) = 0\,. \tag{2.8}$$

Dieses System kann nur iterativ aufgelöst werden. Wir verwenden die Methode von Newton-Raphson. Dazu werden die zweiten Ableitungen der Likelihoodfunktion nach den Parametern benötigt. Diese bilden eine Matrix F. Es ist jedoch in solchen Fällen üblich, an Stelle von F die Matrix $I = -E(F)$ zu verwenden. Die Matrix I heisst Informationsmatrix; ihre Elemente sind Linearkombinationen der Gewichte W_{ij}.

$$W_{ij} = \frac{N_{ij}}{\pi_{ij}(1-\pi_{ij})} \left(\frac{\partial \pi_{ij}}{\partial z_{ij}}\right)^2 . \tag{2.9}$$

Mit den Definitionen

$$\overrightarrow{\mathrm{Par}} = (\beta_1, \beta_2, \ldots, \beta_b, \alpha, \gamma)^T \text{ und}$$

$$\vec{\phi} = \left(\frac{\partial \ln L}{\partial \beta_1}, \ldots, \frac{\partial \ln L}{\partial \gamma}\right)^T$$

kann das Iterationsverfahren übersichtlich dargestellt werden. Die Gleichung $\vec{\phi} = 0$ wird ersetzt durch

$$\underline{\vec{\phi}^{(0)} - I^{(0)} \cdot \delta\, \overrightarrow{\mathrm{Par}} = 0} \tag{2.10}$$

oder

$$\delta\, \overrightarrow{\mathrm{Par}} = (I^{(0)})^{-1}\, \vec{\phi}^{(0)} . \tag{2.11}$$

I ist eine $(b+2) \cdot (b+2)$-Matrix; sie wächst mit der Zahl der Parameter. I ist jedoch von einfacher Struktur, so dass das durch (2.10) dargestellte lineare Gleichungssystem ohne grossen Aufwand zu lösen ist. Die inverse Matrix I^{-1} wird bei der Berechnung der Testgrössen nicht benötigt. Es genügt, die zu α und γ gehörenden Elemente von I^{-1} zu kennen.

Mit den Definitionen

$$W_i = W_{i1} + W_{i2} , \tag{2.12}$$

$$W_{i\alpha} = (W_{i1} - W_{i2}) \cdot \delta_{i1}^{(\alpha)} , \tag{2.13}$$

$$W_{i\gamma} = - W_{i1} + W_{i2} , \tag{2.14}$$

$$W_{\alpha\gamma} = \sum_i \sum_j \delta_{ij}^{(\alpha)} (-1)^j W_{ij} , \tag{2.15}$$

$$W_{\alpha\alpha} = W_{\gamma\gamma} = \sum_i W_i . \tag{2.16}$$

kann I folgendermassen dargestellt werden:

$$I = \left[\begin{array}{ccc|cc} W_{1.} & & 0 & W_{1\alpha} & W_{1\gamma} \\ & \cdot & & \cdot & \cdot \\ & & \cdot & \cdot & \cdot \\ & & \cdot & \cdot & \cdot \\ 0 & & \cdot & \cdot & \cdot \\ & & W_{b.} & W_{t\alpha} & W_{b\gamma} \\ \hline W_{1\alpha} & \ldots & W_{l\alpha} & W_{\alpha\alpha} & W_{\alpha\gamma} \\ W_{1\gamma} & \ldots & W_{l\gamma} & W_{\alpha\gamma} & W_{\gamma\gamma} \end{array}\right]. \tag{2.17}$$

Die Gleichung (2.10) stellt ein System von $b + 2$ linearen Gleichungen dar. Die ersten b Gleichungen werden nach $\delta\beta_i$ aufgelöst:

$$\delta\beta_i = \frac{\phi_{\beta i}}{W_{i.}} - \frac{W_{i\alpha}}{W_{i.}}\,\delta\alpha - \frac{W_{i\gamma}}{W_{i.}}\,\delta\gamma\,. \tag{2.18}$$

Die $\delta\beta_i$ werden in die letzten beiden Gleichungen eingesetzt; man erhält:

$$\left.\begin{array}{l} W'_{\alpha\alpha}\,\delta\alpha + W'_{\alpha\gamma}\,\delta\gamma = \phi'_\alpha \\[2ex] W'_{\alpha\gamma}\,\delta\alpha + W'_{\gamma\gamma}\,\delta\gamma = \phi'_\gamma \end{array}\right\} \tag{2.19}$$

mit den Definitionen

$$W'_{\alpha\alpha} = W_{\alpha\alpha} - \sum_i \frac{W^2_{i\alpha}}{W_{i.}} = W'_{\gamma\gamma}\,, \tag{2.20}$$

$$W'_{\alpha\gamma} = W_{\alpha\gamma} - \sum_i \frac{W_{i\alpha}\cdot W_{i\gamma}}{W_{i.}}\,, \tag{2.21}$$

$$\phi'_\alpha = \phi_\alpha - \sum_i \frac{W_{i\alpha}}{W_{i.}}\,\phi_{\beta i}\,, \tag{2.22}$$

$$\phi'_\gamma = \phi_\gamma - \sum_i \frac{W_{i\gamma}}{W_{i.}}\,\phi_{\beta i}\,. \tag{2.23}$$

(2.19) wird nach dem üblichen Eliminationsverfahren aufgelöst; man erhält für $\delta\alpha$ und $\delta\gamma$:

$$\delta\alpha = \frac{W'_{\alpha\chi}\,\phi'_\alpha - W'_{\alpha\gamma}\cdot\phi'_\gamma}{(W'_{\alpha\alpha})^2 - (W'_{\alpha\gamma})^2}\,, \tag{2.24}$$

$$\delta\gamma = \frac{W'_{\alpha\alpha}\,\phi'_\gamma - W'_{\alpha\gamma}\,\phi'_\alpha}{(W'_{\alpha\alpha})^2 - (W'_{\alpha\gamma})^2}\,. \tag{2.25}$$

Die $\delta\beta_i$ werden mit (2.19) berechnet. Nach dem Iterationsschritt erhält man als neue Parameterwerte:

$$\hat{\beta}_i^{(1)} = \hat{\beta}_i^{(0)} + \delta\beta_i \,, \tag{2.26}$$

$$\hat{\alpha}^{(1)} = \hat{\alpha}^{(0)} + \delta\alpha \,, \tag{2.27}$$

$$\hat{\gamma}^{(1)} = \hat{\gamma}^{(0)} + \delta\gamma \,. \tag{2.28}$$

Die Iteration wird abgebrochen, sobald die Beträge der Änderungen kleiner als eine frei gewählte Grenze ε geworden sind.

Besonders einfach wird die Analyse, wenn für alle i gilt: $N_{i1} = N_{i2}$ und Winkeltransformation verwendet wird. In diesem Falle sind die β_i mit α und γ nicht korreliert.

3. **Tests**

Die Streuungszerlegung des üblichen Cross-over-Versuches kann nicht direkt übernommen werden, weil eine Zerlegung in die Teile «Zwischen Patienten» und «Innerhalb Patienten» im allgemeinen nicht mehr möglich ist. Eine Ausnahme bildet der am Ende von Kapitel 2 erwähnte Fall.

Wir versuchen Testgrössen für die Verfahren und die Versuchsperioden zu finden. Zuerst muss aber geprüft werden, ob das gewählte Modell den Daten entspricht. Zu diesem Zwecke wird ein χ^2-Anpassungstest mit $b - 2$ Freiheitsgraden durchgeführt.

Zur χ^2-Testgrösse gelangt man auch durch Betrachten von d_{ij}. Diese Grösse wird 0, wenn $N_{ij} \cdot \pi_{ij} = x_{ij}$. Sie misst also Abweichungen des Modellwertes vom gemessenen Wert. Für d_{ij} gilt:

$$E(d_{ij}) = 0 \quad V(d_{ij}) = \frac{N_{ij}}{\pi_{ij}(1 - \pi_{ij})} \left(\frac{\partial \pi_{ij}}{\partial z_{ij}} \right)^2 = W_{ij} \,. \tag{3.1}$$

Im Grenzfall grosser N_{ij} strebt d_{ij}^2 / W_{ij} gegen eine Chiquadratverteilung mit einem Freiheitsgrad. Die Summe unabhängiger χ^2-Grössen ist wiederum wie χ^2 verteilt. Da aber für π_{ij} nur Schätzwerte $\hat{\pi}_{ij}$ zur Verfügung stehen, muss die Gesamtzahl der Freiheitsgrade von $2\,b$ um $b + 2$ auf $b - 2$ reduziert werden. Falls $\chi^2 > \chi^2_\alpha$, so ist das gewählte lineare Modell zu verwerfen.

Zur Berechnung von Testgrössen für die Hypothesen

1) $\alpha = 0$ und $\gamma = 0$,

2) $\alpha = 0$,

3) $\gamma = 0$

benötigen wir die Kovarianzmatrix, das heisst die inverse Informationsmatrix. Es genügt jedoch, den zu α und γ gehörenden Teil $\Sigma_{\alpha\gamma}$ zu kennen. Es lässt sich leicht zeigen (siehe Kapitel 2), dass gilt:

$$\sum_{\alpha\gamma} = \begin{pmatrix} W'_{\alpha\alpha} & W'_{\alpha\gamma} \\ W'_{\alpha\gamma} & W'_{\alpha\alpha} \end{pmatrix}^{-1} = \frac{1}{(W'_{\alpha\alpha})^2 - (W'_{\alpha\gamma})^2} \begin{pmatrix} W'_{\alpha\alpha} & -W'_{\alpha\gamma} \\ -W'_{\alpha\gamma} & W'_{\alpha\alpha} \end{pmatrix}$$

$$= \begin{pmatrix} V(\hat{\alpha}) & \mathrm{Cov}(\hat{\alpha}, \hat{\gamma}) \\ \mathrm{Cov}(\hat{\alpha}, \hat{\gamma}) & V(\hat{\gamma}) \end{pmatrix}. \tag{3.2}$$

Die folgenden Testgrössen sind dann asymptotisch wie χ^2 verteilt:

1) Für $\alpha = 0$ und $\gamma = 0$: $\chi^2_{\alpha\gamma} = (\hat{\alpha}, \hat{\gamma}) \sum_{\alpha\gamma}^{-1} \begin{pmatrix} \hat{\alpha} \\ \hat{\gamma} \end{pmatrix}$ FG $= 2$. (3.3)

2) Für $\alpha = 0$: $\chi^2_{\alpha} = \dfrac{\hat{\alpha}^2}{V(\hat{\alpha})}$ FG $= 1$. (3.4)

3) Für $\gamma = 0$: $\chi^2_{\gamma} = \dfrac{\hat{\gamma}^2}{V(\hat{\gamma})}$ FG $= 1$. (3.5)

Wir fassen die Testgrössen wie bei der Streuungszerlegung üblich in einer Tabelle zusammen:

Streuung	FG	$\underset{\sim}{SQ} = \chi^2$
Behandlung, bereinigt	1	χ^2_{α}
Perioden, bereinigt	1	χ^2_{γ}
Behandlung und Perioden, bereinigt	2	$\chi^2_{\alpha\gamma}$
Rest (Anpassung)	$b - 2$	$\sum_{i,j} \dfrac{(x_{ij} - N_{ij}\,\hat{\pi}_{ij})^2}{N_{ij}\,\hat{\pi}_{ij}(1 - \hat{\pi}_{ij})}$

4. **Beispiel**

Ein Kopfwehmittel ist in einem Doppelblindversuch an 21 Patienten geprüft worden. Während der ersten Versuchsperiode von etwa 60 Tagen haben 9 Patienten ein Placebo und 11 Patienten ein Testpräparat erhalten. Während der zweiten Versuchsperiode wird die Zuteilung umgekehrt. Die Versuchsdauer variiert zwischen 42 und 62 Tagen pro Periode. Als Mass für die Wirksamkeit dient der Anteil an Kopfwehtagen. In Tabelle 1 sind die Zahl der beobachteten Tage sowie die Kopfwehtage zusammengestellt.

Tabelle 1
Versuchsdauer und Kopfwehtage.

Reihenfolge	Patient	1. Periode		2. Periode	
		beob.	mit Anf.	beob.	mit Anf.
Placebo → Testpräparat	1	57	39	57	35
	2	57	55	42	42
	3	47	40	57	41
	4	57	53	57	46
	5	48	17	57	15
	6	57	44	57	40
	7	57	42	57	35
	8	57	56	57	57
	9	56	56	57	56
Testpräparat → Placebo	1	57	9	64	18
	2	57	45	53	49
	3	58	57	57	57
	4	57	57	45	42
	5	57	55	57	57
	6	57	28	42	19
	7	62	46	57	47
	8	57	15	53	19
	9	59	57	57	57
	10	58	25	57	28
	11	53	35	57	42

Bei einer ersten Analyse des Datenmaterials werden Unterschiede in der Beobachtungsdauer vernachlässigt. Die Prozentzahlen werden mittels $z = \arcsin\sqrt{\pi}$ in Winkelgrade übergeführt; mit den Werten z wird sodann die übliche Streuungszerlegung durchgeführt. Wir beschränken uns auf die Analyse «innerhalb der Patienten» und erhalten:

Tabelle 2
Streuungszerlegung der Werte $z = \arcsin\sqrt{\pi}$.

Streuung	FG	SQ	F
Testpräparat, bereinigt	1	0,050226	5,02*
Periode, bereinigt	1	0,003089	0,34
Testpräparat und Periode	2	0,053315	3,06
Rest	18	0,157007	–

Die Wirkung des Testpräparates ist auf dem 5%-Niveau gesichert.

Um die in den Kapiteln 2 und 3 dargelegte Theorie anwenden zu können, muss eine geeignete Transformation gesucht werden. Wir nehmen an, die Kopfwehanfälle seien nach Poisson mit dem Parameter λ verteilt, unter der Wirkung des Testpräparates jedoch mit dem Parameter $\alpha\,\lambda$. Die Wirkung ist linear in der Logarithmusdarstellung

$$\ln(\alpha\,\lambda) = \ln\alpha + \ln\lambda\,.$$

Die Wahrscheinlichkeit für einen Tag mit Kopfwehanfällen ist

$$\pi = \sum_{x=1}^{\infty} \frac{e^{-\alpha\lambda} \cdot (\alpha\,\lambda)^x}{x!}$$

oder

$$\pi = 1 - e^{-\alpha\lambda}.$$

Dann gilt $1 - \pi = e^{-\alpha\lambda}$.

Zweimaliges Logarithmieren führt zur linearen Darstellung in den Parametern:

$$\ln\big(-\ln(1-\pi)\big) = \ln\alpha + \ln\lambda\,.$$

Dies ist genau die komplementäre Loglog-Transformation. Sie ist zum ersten Mal von Mather [4] angegeben worden.

Die Anfangswerte für das Iterationsverfahren werden nach der Methode der kleinsten Quadrate aus den direkt transformierten Prozentwerten berechnet. In Extremfällen wird 0% durch 0,1% und 100% durch 99,9% ersetzt; auf diese Weise werden unendlich grosse Werte, wie sie bei der Probit-, Logit- und komplementären Loglog-Transformation vorkommen, vermieden.

Patienten mit den Wertepaaren 0%/0% und 100%/100% werden von der Auswertung ausgeschlossen; die Iteration kann wegen $|z| \to \infty$ nicht konvergieren.

Mit den Zahlen aus Tabelle 1 erreicht man Stabilität schon nach fünf Iterationsschritten bei einer Schranke $\varepsilon = 0{,}0001$. Wir erhalten folgende Resultate:

$$\hat{\alpha} = -0{,}122\,,$$

$$\hat{\gamma} = 0{,}001\,,$$

$$\sum\nolimits_{\alpha\gamma} = \begin{pmatrix} 0{,}001\,081 & 0{,}000\,037 \\ 0{,}000\,037 & 0{,}001\,081 \end{pmatrix}.$$

Die Testgrössen werden in einer Tabelle zusammengefasst.

Tabelle 3
Testgrössen.

Streuung	FG	$SQ = \chi^2$	$\chi^2_{0.05}$
Testpräparat, bereinigt	1	13,668*	3,841
Periode, bereinigt	1	0,001	3,841
Testpräparat und Periode, bereinigt	2	13,692*	5,991
Rest (Anpassung)	18	21,528	28,869

Die komplementäre Loglog-Transformation darf als eine geeignete Transformation angesehen werden. Das Testpräparat ergibt einen deutlich signifikanten Effekt; eine Wirkung der Versuchsperiode kann jedoch nicht beobachtet werden. Der Effekt des Testpräparates ist deutlicher als bei der Analyse mit Winkelgraden.

Literatur

[1] Berchtold, W. und Linder, A. (1973): Streuungszerlegung mit Anteilziffern. EDV in Medizin und Biologie *4*, Heft 3, 99–108.

[2] Cochran, W. G. and Cox, G. M. (1966): Experimental Designs. J. Wiley, New York.

[3] Linder, A. (1964): Statistische Methoden für Naturwissenschafter, Mediziner und Ingenieure. 4. Aufl. Birkhäuser, Basel.

[4] Mather, K. (1949): The analysis of extinction time data in bioassay. Biometrics *5*, 127–143.

Adresse des Autors:
Willi Berchtold, Laboratorium für Biometrie und Populationsgenetik, ETH Zürich, Gloriastrasse 35, CH-8006 Zürich.

Aurelio Palazzi

Aurelio Palazzi

Nonparametric Analysis of Variance – Tables of Critical Values

Summary

The nonparametric methods of comparing series of observations allow statistical inference made after some very simple arithmetical operations.

No stringent limitation is to be made as far as the nature of the parent population, this can also be of unknown shape. Particularly interesting is the application of this method to the analysis of t series of observations (with n replications each) either random or within blocks.

For the two cases tables are given of the critical value of the sums of the squares of the rank subtotals.

The tables range from $t = 3 \div 6$ and $n = 3 \div 20$ and are calculated for the probability levels $P = 90, 95, 99$.

1. Introduction

In situations where the dependent variate of a designed experiment can be expressed as a numerical continuous value and the parent distribution can be considered normal, the quantitative evaluation of the data by plain application of exact methods such as analysis of variance is a straightforward procedure.

Often in practice the scientist is faced with situations where the data can be expressed only by code numbers and/or the underlying more or less known parent distribution is clearly (or suspected to be) far from the normal model.

In this case it is therefore profitable to have recourse to rapid evaluation methods based on the ranked succession of data arranged in increasing or decreasing order (rank methods) which only call for few and very simple arithmetical operations and the use of ready calculated numerical tables.

The methods based on the ranking of observations do not depend on any particular distribution model; for their practical application it is enough that the distribution should be continuous.

Where the experiment may lead to data necessarily expressed in the form of a classification, no exact method can have higher efficiency; therefore here rank methods have no true competitors.

Experiments with normally distributed quantitative dependent variate can also be evaluated by rank methods simply by converting numerical values back into ranks, keeping the individuality of actual blocks.

In the latter cases advantages in computation are obtained at the expense of efficiency: but for usual practical applications this is not a great inconvenience although rank methods cannot be considered as real and proper alternatives for the exact methods especially in border-line cases.

Today it is recognized that nonparametric methods contain a great potentiality of didactic value for non professional statisticians and students of sciences.

In many schools the illustration of nonparametric methods is made before entering the complicated world of the parametric approach.

2. Comparison of Several Series of Observations in a One-Classification Scheme

It is here intended to compare t treatments, each replicated n times, all the $t\,n$ data are listed in increasing (or decreasing) order and each (of the $t\,n$) datum is given a rank.

In each treatment series the addition of the ranks is made and then the sum of the squares of each of such subtotals is effected.

Let $\Sigma\, S_t^2$ be the value under consideration; this is to be compared in the usual way with the critical values shown in Table 1 for the three probability levels $P = 90; 95; 99$.

If $\Sigma\, S_t^2$ s bigger than or equal to the critical value at $P = 99$, the conclusion is reached that the difference between treatments is significant.

The table is based on the formula:

$$\chi^2 = \frac{12\, \Sigma\, S_t^2}{t\, n^2(t\, n + 1)} - 3\,(t\, n + 1)\,,$$

where χ^2 stands for the known probabilistic function for $t - 1$ degrees of freedom and for the assumed levels of probability, all other symbols being known.

The expression of the above formula in terms of χ^2 is a satisfying approximation in most practical cases, especially with increasing t and n.

Table 1
Critical values of the total sum of the squares of the rank totals for each of the t series ($\Sigma_t^2 S$) in the case of comparison of t series of observations, using a one-criterion classification (t series each comprising r random replications; the rank is related to $t\,r$ observations).

	$P = 90$				95				99			
	$t = 3$	4	5	6	3	4	5	6	3	4	5	6
$n = 3$	779	1 765	3 347	5 663	810	1 826	3 449	5 820	882	1 963	3 677	6 163
4	2 267	5 191	9 909	16 847	2 340	5 333	10 148	17 214	2 507	5 652	10 679	18 017
5	5 260	12 119	23 232	39 616	5 399	12 393	23 695	40 327	5 721	13 010	24 721	41 883
6	10 534	24 375	46 862	80 077	10 771	24 844	47 657	81 299	11 322	25 902	49 419	83 973
7	19 028	44 170	85 098	145 632	19 402	44 911	86 354	147 564	20 269	46 581	89 139	151 795
8	31 842	74 897	142 985	244 978	32 396	75 198	144 854	247 854	33 684	77 680	148 996	254 151
9	50 239	117 134	226 322	388 111	51 025	118 696	228 975	392 196	52 650	122 219	234 858	401 142
10	75 644	176 643	341 655	586 320	76 918	178 780	345 287	591 914	79 213	183 599	353 339	604 162
11	109 643	256 371	496 283	852 192	111 069	259 209	501 108	859 626	114 379	265 609	511 805	875 905
12	153 986	360 446	698 251	1 199 608	155 832	364 125	704 506	1 209 248	160 120	362 418	718 374	1 230 356
13	210 582	493 384	956 358	1 643 748	212 925	498 054	964 300	1 655 991	218 365	508 581	981 910	1 682 800
14	281 506	660 033	1 280 150	2 201 086	284 427	665 907	1 290 060	2 216 363	291 208	679 038	1 312 029	2 249 816
15	363 990	865 823	1 679 925	2 889 391	372 577	872 979	1 692 102	2 908 167	380 906	689 110	1 719 099	2 949 280
16	475 433	1 116 272	2 166 731	3 727 730	479 780	1 124 947	2 181 496	3 750 501	489 875	1 144 505	2 214 233	3 800 364
17	603 393	1 417 479	2 752 363	4 736 466	608 600	1 427 875	2 770 061	4 763 762	620 694	1 451 313	2 809 299	4 823 534
18	755 590	1 775 879	3 449 370	5 937 255	761 765	1 788 209	3 470 365	5 969 640	776 106	1 816 008	3 516 913	6 040 555
19	934 908	2 198 289	4 271 049	7 353 054	942 163	2 212 780	4 295 727	7 391 123	959 013	2 245 451	4 350 440	7 474 485
20	1 144 390	2 691 911	5 231 446	9 008 111	1 152 045	2 708 802	5 760 215	9 052 494	1 172 481	2 746 883	5 323 996	9 149 681

Table 2 shows (as an example of practical application) the data referring to the soundness of the surface apperaence of cold rolled steel samples that may, in more or less marked way, bear traces of stretches due to Bauschinger effect; that at three different levels of the content of interstitial trace elements: low, medium and high (concentration by weight).

Both variables are illdefined, the independent one because the chemistry is only a faded indication of the true interstitial situation, the dependent one because it can only be qualitatively appreciated as a visual combination of intensity and counting: a typical situation for a non parametric analysis.

To each level of interstitials correspond ten independent samples; this experiment is evidently of the one-classification type. Considering all the data as if they belonged to one set, the thirty observations are given the ranks (increasing with growing defective appearance).

A sum of the squares ΣS_t^2 amounting to 83749 is calculated; the critical value for $P = 99$, $t = 3$ and $n = 10$ is shown in Table 1 as 79213, which is considerably lower than the value found; it is therefore concluded that the difference between the three series is significant: by increasing interstitials, stretching significantly increases, i.e. surface quality gets worse.

Table 2
Ranking for defectiveness (stretching due to Bauschinger effect) related to different interstitial content[1]).

	Interstitial content			
	Low	Medium	High	
	4.5	21	29	
	12	13	20	
	1	14	11	
	7	24	30	
	17	18	16	
	10	19	23	
	8	22	26	
	2	4.5	27	
	9	25	15	
	3	6	28	
Subtotals S_t	73.5	166.5	225	$\Sigma S_t^2 = 83749.50$

[1]) Fractional figures are ties.

3. Comparison of Several Series of Observations in a Two-Classification Scheme

It is now intended to compare t treatments each being replicated n times in n ordered series of replications.

We begin by arranging the t data (within each replication) in an increasing (or decreasing) order and then each t datum is given a corresponding rank. In each treatment series the sum of the ranks is calculated and then the sum of the squares of each of such t subtotals is also calculated.

Let $\Sigma S_t'^2$ be the described value: this is compared in the usual way with the critical values listed in Table 3 for the three probability levels $P = 90$, 95, 99.

If $\Sigma S_t'^2$ is bigger than, or equal to, the critical value for $P = 99$, it is to be concluded that the difference is significant. The table is based on the formula:

$$\chi^2 = \frac{12 \Sigma S_t'^2}{t\, n(t+1)} - 3\, n(t+1)$$

with $t - 1$ degrees of freedom, for which the limitations and the cautions suggested for the one-classification case still hold.

As an example of practical application Table 4 shows data referring to the heat erosion undergone by refractory bricks of 4 different brands here marked A B C D.

The erosion was evaluated by the physical aspect of the bricks after the dipping test, which was carried out by putting groups of $t = 4$ bricks (one for each of the four brands to be compared) into the liquid slag above the molten steel.

The results obtained from such tests are typically ones to be evaluated by rank methods and consequently their use is the most efficient since an 'exact' method of evaluation cannot be defined.

The design of this experiment evidently belongs to the two-classification type (random blocks). In fact the bricks of the 4 different brands to be compared were dipped together, that is at the same time, in the same slag, and tests so structured were made for $n = 20$ heats.

After the bricks were lifted out they were numbered 1, 2, 3, 4 according to their erosion depth, the smaller number corresponding to the slightest erosion and therefore to the best brick quality.

A sum of squares $\Sigma S_t'^2$ equal to 10604, is calculated; the critical value for $P = 99$, $t = 4$ and $n = 20$ is shown in Table 3 as 10274, clearly lower than the value found.

It is then concluded that the 4 brands of bricks significantly differ from one another because in the dipping test they behaved increasingly better in accordance with the succession A B C D.

The use of a two-classification experiment (brands and heats) has made it possible to free the comparison from the block effect which might be so high as to upset, in some cases, the expected result owing to the sudden arising of unexpected sources of variations.

Table 3
Critical values of the sum of the squares of the rank totals for each t series ($\Sigma S'^2_t$) in the case of the comparison of t series of observations in a two-criterion classification (t series each with n ordered replications). The rank is related to t observations within each block of replications.

	$P = 90$				95				99			
	$t = 3$	4	5	6	3	4	5	6	3	4	5	6
$n = 3$	122	256	432	722	126	264	444	741	136	282	471	781
4	210	398	798	1 305	216	407	815	1 331	229	428	853	1 387
5	323	650	1 173	1 942	330	663	1 194	1 973	346	691	1 239	2 041
6	460	963	1 737	2 840	468	978	1 762	2 878	487	1 013	1 819	2 963
7	620	1 224	2 274	3 750	630	1 241	2 303	3 794	652	1 280	2 368	3 980
8	805	1 641	3 036	4 963	816	1 662	3 070	5 014	842	1 707	3 146	5 126
9	1 013	2 119	3 735	6 145	1 026	2 142	3 773	6 202	1 055	2 195	3 856	6 327
10	1 246	2 500	4 694	7 673	1 260	2 525	4 737	7 737	1 292	2 581	4 832	7 878
11	1 503	3 083	5 556	9 129	1 518	3 111	5 602	9 199	1 553	3 174	5 704	9 351
12	1 783	3 725	6 713	10 972	1 800	3 756	6 765	11 049	1 839	3 827	6 878	11 218
13	2 088	4 226	7 737	12 701	2 106	4 259	7 792	12 783	2 148	4 333	7 913	12 964
14	2 416	4 974	9 092	14 859	2 436	5 010	9 152	14 948	2 481	5 091	9 285	15 145
15	2 769	5 781	10 278	16 860	2 790	5 020	10 341	16 956	2 833	5 909	10 481	17 164
16	3 146	6 403	11 831	19 333	3 168	6 443	11 900	19 436	3 219	6 535	12 051	19 661
17	3 546	7 315	13 179	21 608	3 570	7 359	13 250	21 716	3 625	7 458	13 410	21 953
18	3 971	8 288	14 930	24 396	3 996	8 334	15 007	21 511	4 054	8 440	15 177	24 764
19	4 419	9 029	16 440	26 944	4 446	9 077	16 520	27 065	4 507	9 187	16 698	27 330
20	4 892	10 106	18 389	30 047	4 920	10 158	18 474	30 175	4 984	10 274	18 664	30 456

Table 4
Rank classification related to the erosion depth during the dipping test of bricks for steel ladles (lower number, slighter erosion, that is better quality).

Test (Heat)	Brand				
	A	B	C	D	
1	4	2.5	1	2.5	
2	4	2.5	2.5	1	
3	4	3	2	1	
4	3.5	2	3.5	1	
5	4	2.5	2.5	1	
6	1	3	4	2	
7	3	4	2	2	
8	4	3	1.5	1.5	
9	3.5	2	1	3.5	
10	4	1	2	3	
11	4	3	2	1	
12	2	4	2	2	
13	2.5	1	2.5	4	
14	4	2	3	1	
15	3	4	2	1	
16	2	1	4	3	
17	4	1.5	3	1.5	
18	4	3	2	1	
19	4	3	2	1	
20	4	2	3	1	
Subtotals S'_t	68.5	50.0	47.5	34.0	$\Sigma S_t'^2$ 10 604

References

[1] TUKEY, J. W. (1953): Some Selected Quick and Easy Methods of Statistical Analysis. New York Acad. Sci., Ser. II, Vol. *16*, No. 2, 88–97.
[2] WALLIS, W. A. (1952): Rough-and-ready Statistical Tests. Ind. Qual. Control, 35–40.
[3] WILCOXON, F. (1949): Some Rapid Approximate Statistical Procedures. American Cyanamid Co., Stamford, Conn.

Author's address:
Aurelio Palazzi, Centro Sperimentale Metallurgico, Rome, Italy.

Henri L. Le Roy

Henri L. Le Roy

A Special Use of the General Interpretation for the One-Way Analysis of Variance in Population Genetics of Quantitative Characters

1. Introduction

Suppose we have sampled numerical values of the character X from N individuals. These N individuals are generated by w females (dams) and s males (sires), as illustrated in Fig. 1.

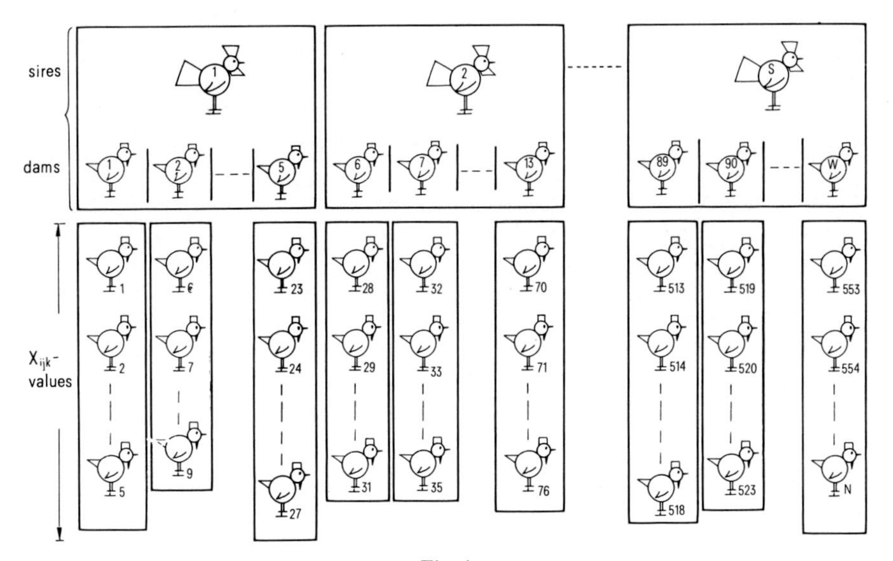

Fig. 1

Animals in the same dam-group are full-sisters, because they have identical parents. Individuals belonging to the same sire-group but different dam-groups are paternal half-sisters; these animals have the same father (sire) but different mothers (dams).

The analysis of variance which is used for estimating the population parameter h^2 (coefficient of heritability) is shown in Table 1.

Table 1
The analysis of variance for the values characterised in Fig. 1.

Source of variation	*df*	*MS*
between sire-groups	$s-1$	$MS(S)$
between dam-groups within sire-groups	$w-s$	$MS(D)$
between full-sisters (within dam-groups)	$N-w$	$MS(W)$

If the number of sire-groups is relatively small, the confidence interval, related to the estimated h^2, using the sire component as an estimate of the genetic variance, has a range which is too large and therefore the use of h^2 for practical application is questionable. In this case, we use a combined estimate of heritability, calculated from the sire *and* dam component of variance, namely h^2_{D+S}, which has a smaller confidence interval.—It will be shown that it is possible to construct a combined estimate of h^2_{D+S}, symbolised by ${}_c h^2_{D+S}$, using the one-way analysis of variance, mentioned in Table 2.

Table 2
The one-way analysis of variance used for estimate h^2.

Source of variation	*df*	*SS*	*MS*	$E[MS]$
Between dam-groups	$w-1$	$SS(D) = B - C$	$MS(D) = \dfrac{B-C}{w-1}$	$\sigma_2^2 + k\,\sigma_1^2$
Within dam-groups	$N-w$	$SS(W) = A - B$	$MS(W) = \dfrac{A-B}{N-w}$	σ_2^2

$A = \Sigma_i\, \Sigma_j\, \Sigma_k\, X^2_{ijk},\ B = \Sigma_i\, \Sigma_j\, X^2_{ij.}/N_{ij}$ and $C = X^2.../N$
$\Sigma_i\, \Sigma_j\, \Sigma_k(1) = N,\ \Sigma_i\, \Sigma_j(1) = w$ and $\Sigma_i(1) = s$

i: indices fot the i-th sire-group
ij: indices for the j-th dam-group within the i-th sire-group
ijk: the k-th offspring within the ij-th dam-group

The problem is to find the expected structure of σ_2^2 and σ_1^2 and also to construct the coefficient of heritability which should be as near as possible to the ratio σ^2_{Ga}/σ^2_P. To explain the genetical part of the problem, we consider the following model for the individual measurement:

$X = P = \mu + ge + u$ as the simplest interpretation

- u → environmental deviation
- ge → genotypic deviation
- μ → general mean
- P → phenotypic value = realised measurement for the character X

It is easily seen that $\sigma^2_P = \sigma^2_{Ge} + \sigma^2_U$, if $r_{Ge.U} = 0$.

An improved model is given by

$$X = P = \mu + ga + do + u_L + u_R$$

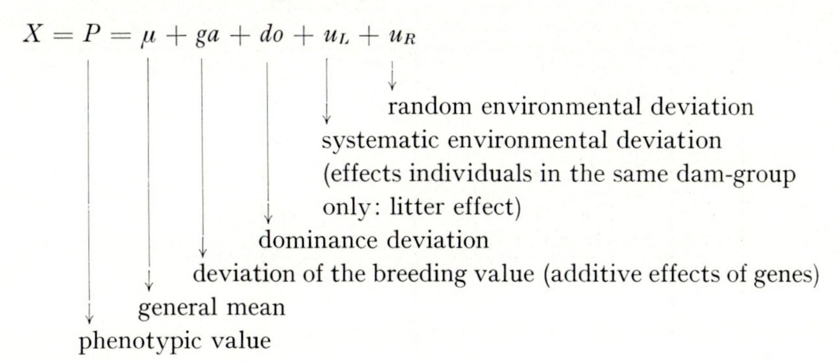

The following structure of σ_P^2 will be found true:

$$\sigma_P^2 = \underbrace{\sigma_{Ga}^2 + \sigma_{Do}^2}_{\sigma^2_{Ge}} + \underbrace{\sigma_{UL}^2 + \sigma_{UR}^2}_{\sigma^2_U}$$

Using the improved model, it can be shown that for a random mated population the covariances between individual measurements are given by (Kempthorne [1], Le Roy [2]):

$\text{Cov}(FS) = \sigma_{Ga}^2/2 + \sigma_{Do}^2/4 + \sigma_{UL}^2$	FS:	full-sibs,
$\text{Cov}(HS) = \sigma_{Ga}^2/4$	HS:	half-sibs,
$\text{Cov}(NR) = 0$	NR:	not related measurements (animals).

2. The Expectation of $MS(D)$ and $MS(W)$ resp. the Expectation of $E[\sigma_2^2]$, $E[\sigma_1^2]$ and $E[ch_{D+S}^2]$

The structure in the list of observed data can be symbolised as follow:

Measurements:

| xxx | xxxx | .... | xxx || .... || xxxx | xxx | .... | xxx |

Group-frequencies:

$\overleftrightarrow{N_{ij}}$

$\longleftarrow N_{i.} \longrightarrow$ $\qquad\qquad$ $\longleftarrow N_{i'.} \longrightarrow$

$\longleftarrow\qquad\qquad N \qquad\qquad\longrightarrow$

To find the interpretation for the value $A = \Sigma_i \Sigma_j \Sigma_k X_{ijk}^2$ respective $B = \Sigma_i \Sigma_j X_{ij}^2/N_{ij}$ respective $C = X^2.../N$, we use Table 3 to characterize the structure by means of the relevant products called type a, b, c and 0, mentioned in Table 3.

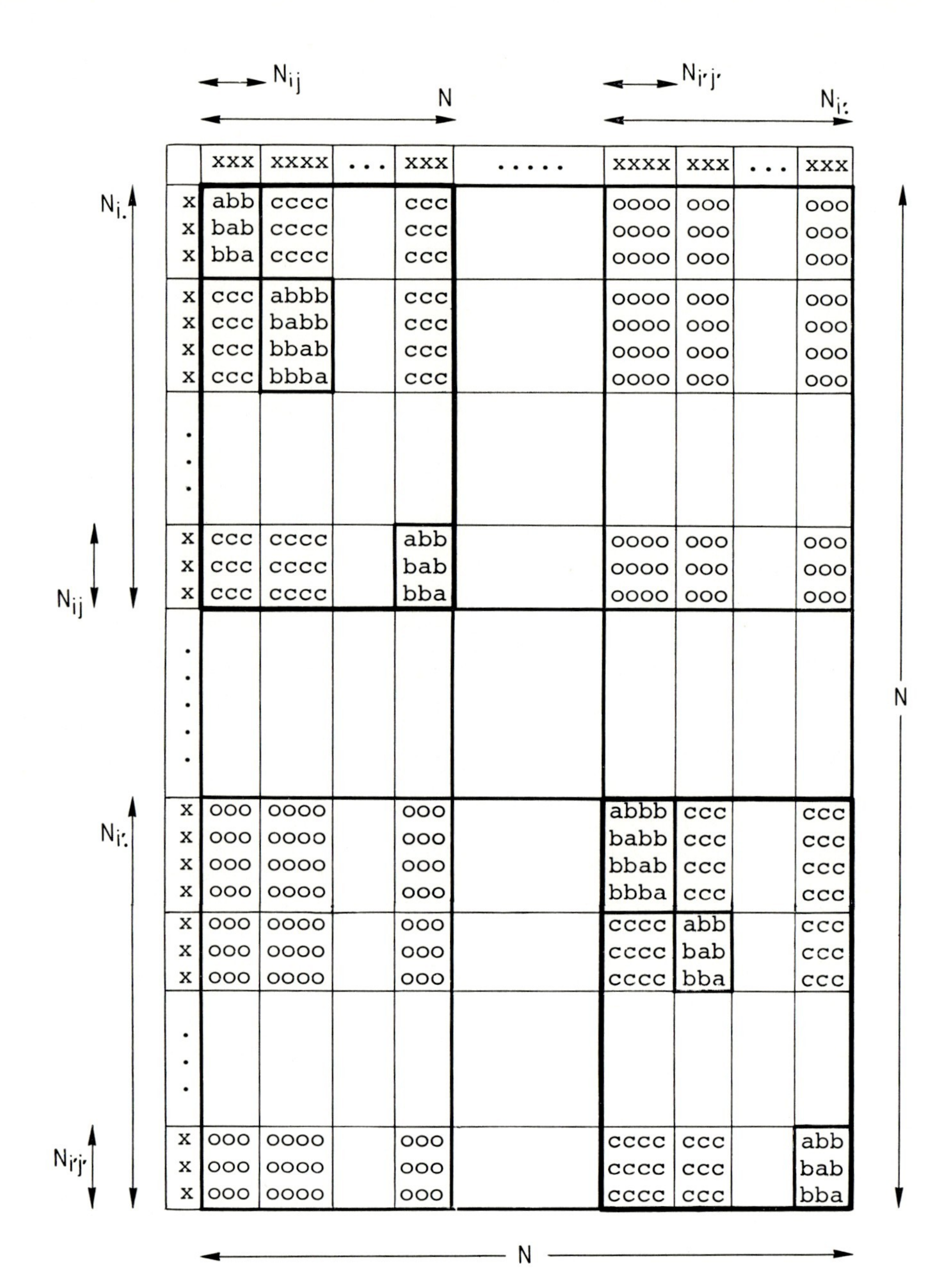

Table 3
The frequencies of the different products between the N measurements X_{ijk}.

Type	Product	Frequency
a	$X_{ijk}\,X_{ijk} = X_{ijk}^2$	N_a
b	$X_{ijk}\,X_{ijk'}\quad k \neq k'$	N_b
c	$X_{ijk}\,X_{ij'k'}\quad j \neq j'$	N_c
0	$X_{ijk}\,X_{i'j'k'}\quad i \neq i'$	N_0

where $N_a + N_b + N_c + N_o = N^2$
and

$$N_a = \Sigma_i\, \Sigma_j\, N_{ij} = N$$
$$N_b = \Sigma_i\, \Sigma_j\, N_{ij}^2 - N$$
$$N_c = \Sigma_i\, N_i^2 - \Sigma_i\, \Sigma_j\, N_{ij}^2$$
$$N_o = N^2 - \Sigma_i\, N_i^2$$

Type b indicates products between measurements for full-sibs.
Type c indicates products between measurements for half-sibs.
Type 0 indicates products between measurements for not related individuals.

The products of type a, b, c and 0 have the following expectations:

Type	Product	Structur	Genetical Interpretation
a	$E[X_{ijk}\, X_{ijk}]$	$\sigma_X^2 + \mu_X^2$	$\sigma_P^2 + \mu^2$
b	$E[X_{ijk}\, X_{ijk'}]_{k \neq k'}$	$\sigma_{kk'} + \mu_X^2$	$\mathrm{Cov}(FS) + \mu^2$
c	$E[X_{ijk}\, X_{ij'k'}]_{j \neq j'}$	$\sigma_{jj'} + \mu_X^2$	$\mathrm{Cov}(HS) + \mu^2$
0	$E[X_{ijk}\, X_{i'j'k}]_{i \neq i'}$	$\sigma_{ii'} + \mu_X^2$	$\mathrm{Cov}(NR) + \mu^2$

Because

$$SS(D) = \Sigma_i\, \Sigma_j\, X_{ij.}^2/N_{ij} - X_{..}^2/N \qquad = B - C\,,$$

$$SS(W) = \Sigma_i\, \Sigma_j\, \Sigma_k\, X_{ijk}^2 - \Sigma_i\, \Sigma_j\, X_{ij.}^2/N_{ij} = A - B\,,$$

the expectations of MS are given by

$$E[MS(D)] = \frac{E[B] - E[C]}{w - 1} \quad \text{and} \quad E[MS(W)] = \frac{E[A] - E[B]}{N - w}\,.$$

Using Table 3, the reader can easily verify that the following expectations hold:

$$\underline{E[A]} = E[\Sigma_i\, \Sigma_j\, \Sigma_k\, X_{ijk}^2] = N \cdot E[a] = N(\sigma_P^2 + \mu^2) = \underline{N\sigma_P^2 + N\,\mu^2}\,, \quad (1)$$

$$\underline{E[B]} = E[\Sigma_i\, \Sigma_j\, X_{ij.}^2/N_{ij}] = \Sigma_i\, \Sigma_j \frac{1}{N_{ij}} E[X_{ij.}^2]$$

$$= \Sigma_i\, \Sigma_j \frac{1}{N_{ij}} E[N_{ij} \cdot a + (N_{ij}^2 - N_{ij})\, b]$$

$$= \Sigma_i\, \Sigma_j \big[(\sigma_P^2 + \mu^2) + [N_{ij} - 1]\big(\mathrm{Cov}(FS) + \mu^2\big)\big]$$

$$= w(\sigma_P^2 + \mu^2) + [N - w]\big(\mathrm{Cov}(FS) + \mu^2\big)$$

$$= \underline{w\,\sigma_P^2 + [N - w]\,\mathrm{Cov}(FS) + N\,\mu^2}\,, \quad (2)$$

$$E[C] = \frac{1}{N} E[X^2_{..}] = \frac{1}{N} \left[N_a \cdot E[a] + N_b \cdot E[b] + N_c \cdot E[c] + N_o \cdot E[o] \right]$$

$$= \frac{1}{N} \big[N(\sigma_P^2 + \mu^2) + \Sigma_i \, \Sigma_j (N_{ij}^2 - N_{ij}) \big(\mathrm{Cov}(FS) + \mu^2\big)$$

$$+ (\Sigma_i \, N_{i.}^2 - \Sigma_i \, \Sigma_j \, N_{ij}^2) \big(\mathrm{Cov}(HS) + \mu^2\big)$$

$$+ (N^2 - \Sigma_i \, N_{i.}^2) \big(\mathrm{Cov}(NR) + \mu^2\big) \big]$$

$$= \underline{\sigma_P^2 + [I^* - 1] \, \mathrm{Cov}(FS) + [II^* - I^*] \, \mathrm{Cov}(HS) + N \mu^2} \, ,$$

$$\mathrm{Cov}(NR) = 0 \, , \tag{3}$$

where $I^* = \Sigma_i \, \Sigma_j \, N_{ij}^2 / N$ and $II^* = \Sigma_i \, N_{i.}^2 / N$.

Simple arithmetic shows that

$$E[MS(D)] = [\sigma_P^2 - \mathrm{Cov}(FS)]$$

$$+ \frac{N - I^*}{w - 1} \left[\mathrm{Cov}(FS) + \frac{I^* - II^*}{N - I^*} \mathrm{Cov}(HS) \right] = \sigma_2^2 + k \, \sigma_1^2$$

$$; \; k = \frac{N - I^*}{w - 1}$$

$$E[MS(W)] = [\sigma_P{}^2 - \mathrm{Cov}(FS)] \qquad = \sigma_2^2$$

Introducing the genetical interpretation of $\mathrm{Cov}(FS)$ and $\mathrm{Cov}(HS)$, we find:

$$\sigma_1^2 = \sigma_{Ga}^2 \, K^{-1} + \frac{1}{4} \sigma_{Do}^2 + \sigma_{UL}^2 \quad \text{with} \quad K^{-1} = \frac{2\,N - I^* - II^*}{4(N - I^*)} \, , \quad \text{and}$$

$$\sigma_2^2 = \frac{1}{2} \sigma_{Ga}^2 + \frac{3}{4} \sigma_{Do}^2 + \sigma_{UR}^2 \, .$$

The best estimate of heritablity is equal to

$$ch_{D+S}^2 = K \, \sigma_1^2 / \left(\sigma_2^2 + \frac{K}{2} \sigma_1^2 \right) = 2 \, K \, \varrho / (2 + K \, \varrho) \, , \quad \text{with} \quad \varrho = \sigma_1^2 / \sigma_2^2$$

with the structure

$$ch^2_{D+S} = \frac{\sigma^2_{Ga} + \left(\frac{K}{4}\right)\sigma^2_{Do} + K.\sigma^2_{UL}}{\sigma^2_{Ga} + \left(0{,}75 + \frac{K}{8}\right)\sigma^2_{Do} + \left(\frac{K}{2}\right)\sigma^2_{UL} + \sigma^2_{UR}}$$

$$\sim \frac{\sigma^2_{Ga} + \left(\frac{K}{4}\right)\sigma^2_{Do} + K.\sigma^2_{UL}}{\sigma^2_P} .$$

The value K is of the following order:

s	w^*	K	
6	4–8	2.14–2.16	
12	4–8	2.07–2.08	w^*: number of dams/sire
24	4–16	2.03	
limit		2	

With regards to the calculated values of MS, the symbol for the components of variance is changed from σ^2 to s^2.

The calculated F-value is identical with

$$F_c = MS(D)_c/MS(W)_c = (s^2_2 + k.s^2_1)/s^2_2 = 1 + k\,R\,, \quad \text{where} \quad R = s^2_1/s^2_2 .$$

Then $R = (1 - F_c)/k$, $k = (N - I^*)/(w - 1)$ and $ch^2_{(D+S)c} = 2\,K\,R/(2 + KR)$, $K = 4(N - I^*)/(2\,N - I^* - II^*)$, see above.

3. The Confidence Interval for $ch^2_{(D+S)c}$

Using the theory for estimating the confidence interval with regard to

$$\varrho = \sigma^2_1/\sigma^2_2\,, \quad \text{we find} \quad \text{Prob}[\varrho_u \leq \varrho \leq \varrho_0] = 1 - \alpha$$

with

$$\varrho_u = [(F_c/F_{\alpha/2;\, w-1,\, N-w}) - 1]/k\,,$$

$$\varrho_0 = [(F_c\,F_{\alpha/2;\, N-w,\, w-1}) - 1]/k\,.$$

Finally the $1-\alpha$ confidence interval for $ch^2_{(D+S)c}$ is given by

$$2\,K\,\varrho_u/(2+K\,\varrho_u) \leq ch^2_{(D+S)} \leq 2\,K\,\varrho_0/(2+K\,\varrho_0)\;.$$

4. Numerical Example (Le Roy [3], pp. 188–191)

Observed datas resp. calculated figures see Table 4.

Computations:

$A = \Sigma_i\,\Sigma_i\,\Sigma_k\,X^2_{ijk} = 2\,063\,984 \quad \text{with} \quad N = 500\;df\,,$

$B = \Sigma_i\,\Sigma_j\,X^2_{ij.}/N_{ij} = 1168^2/16 + 954^2/13 + \cdots + 354^2/5$

$\quad = 1\,939\,028.7 \quad \text{with} \quad w = 59\;df\,,$

$C = (\Sigma_i\,\Sigma_j\,\Sigma_k\,X_{ijk})^2/N = 30825^2/500 = 1\,900\,361.2 \quad \text{with} \quad 1\;df\,.$

Then

$MS(D)_c = [1939\,028.7 - 1900\,361.2]/[59\text{–}1] = 666.681 \quad \text{with} \quad 58\;df\,,$

$MS(W)_c = [2063\,984 - 1939\,028.7]([500\text{–}59] = 283.345 \quad \text{with} \quad 441\;df\,,$

$F_c \quad = 666.681/283.345 = 2.3539$ with $w-1 = 58$ and $N-w$

$\quad = 441\;df\,.$

Furthermore

$I^* \quad = [16^2 + 13^2 + \cdots + 5^2]/500 = 9.26\,,$

$k \quad = [500 - 9.26]/58 = 8.4610\,,$

$R \quad = [2.3529 - 1]/8.4610 = 0.1599\,,$

$II^* = [93^2 + 85^2 + \cdots + 51^2]/500 = 74.02\,,$

$K \quad = 4[500 - 9.26]/[1000 - 9{,}26 - 74.02] = 2{,}14$

Table 4
Shortened table for the observed values. The individual values are not given.

Sire	Dam [1]	Dam [2]	Frequency N_{ij}	Sums $X_{ij.}$ / $X_{i.}$	Sums of squared values
1	1	1	16	$1168 = X_{ij.} = X_{11.}$	$91948 = \Sigma_k X^2_{ijk}$
	2	2	13	954	73908
	3	3	11	770	55616
	4	4	10	691	52471
	5	5	9	615	43135
	6	6	9	679	53557
	7	7	7	454	30312
	8	8	6	451	34263
	9	9	6	368	23578
	10	10	6	346	20782
		10	$93 = N_{1.}$	$6496 = X_{1..}$	$479570 = \Sigma_j \Sigma_k X^2_{1jk}$
2	11	1	14	940	67770
	12	2	14	855	58275
	13	3	13	850	59636
	14	4	10	720	54754
	15	5	10	672	48020
	16	6	9	604	42466
	17	7	8	504	34472
	18	8	7	483	34135
		8	$85 = N_{2.}$	$5628 = X_{2..}$	$399528 = \Sigma_j \Sigma_k X^2_{2jk}$
⋮					⋮
6	44	1	11	659	39962
	45	2	9	628	44512
	46	3	9	502	27162
	47	4	9	639	46737
	48	5	7	375	20739
	49	6	6	335	19355
	50	7	6	311	17907
	51	8	6	320	17856
	52	9	6	357	23799
		9	$69 = N_{6.}$	$4126 = X_{6..}$	$258029 = \Sigma_j \Sigma_k X^2_{6jk}$
7	53	1	11	795	59791
	54	2	10	494	27798
	55	3	8	363	18235
	56	4	7	365	21321
	57	5	5	267	15275
	58	6	5	256	13710
	59	7	5	354	26792
		7	$51 = N_{7.}$	$2894 = X_{7..}$	$182922 = \Sigma_j \Sigma_k X^2_{7jk}$
			$500 = N_{..}$	$30825 = X_{...}$	$2063984 = \Sigma_i \Sigma_j \Sigma_k X^2_{ijk}$

[1] overall number [2] number within sire

and

$$\underline{{}_{c}h^{2}_{(D+S)c}} = 2\,(2.14)\,(0.1599)/[2 + (2,14)\,(0.1599)] = \underline{0.292}\,.$$

Because

$$F_{0,05;\,58.441} = 1.34 \therefore \varrho_u = [(2.35/1.34) - 1]/8.46 = 0.089\,,$$

$$F_{0.05;\,441.58} = 1.44 \therefore \varrho_0 = [(2.35 \cdot 1.44) - 1]/8.46 = 0.282\,.$$

The confidence interval for $\alpha = 0.1$ is then equal to

$$0.174 \leq {}_{c}h^{2}_{(D+S)} \leq 0.464\,.$$

References

[1] KEMPTHORNE, O. (1955): The Theoretical Values of Correlations Between Relatives in Random Mating Populations. Genetics *40*, 153–167.

[2] LE ROY, H. L. (1960): Statistische Methoden der Populationsgenetik. Ein Grundriss für Genetiker, Agronomen und Biomathematiker. Birkhäuser-Verlag, Basel.

[3] LE ROY, H. L. (1966): Elemente der Tierzucht. Genetik-Mathematik-Populationsgenetik. Bayerischer Landwirtschaftsverlag, München.

Author's address:
Henri L. Le Roy, Laboratory of Biometry and Population Genetics, Swiss Federal Institute of Technology, Zurich.

5. Relationships

Pierre Dagnelie

Pierre Dagnelie

L'emploi d'équations de régression simultanées dans le calcul de tables de production forestières

Résumé

L'auteur montre l'intérêt que peut présenter la théorie des équations de régression simultanées pour la construction de tables de production forestières. A cette fin, il étudie trois modèles simples et il expose brièvement les principes de base de l'estimation des paramètres, dans le cas des modèles récursifs et non récursifs.

1. **Introduction**

De longue date, les forestiers européens se sont efforcés de décrire et d'analyser la croissance des peuplements qu'ils avaient à gérer, à l'aide de *tables de production* (Schwappach [11], Wiedemann [12], etc.). Ces tables donnent, en fonction de la qualité et de l'âge des peuplements, des valeurs moyennes relatives aux principales caractéristiques des arbres restant sur pied et prélevés en éclaircie (nombre d'arbres par hectare, hauteur moyenne, circonférence ou diamètre moyen, volume par hectare, accroissement annuel moyen en volume par hectare, etc.).

A l'heure actuelle, les *méthodes de construction* de ces tables doivent évidemment être revues, en fonction notamment des possibilités offertes par la mise en œuvre des moyens modernes de traitement automatique de l'information (Assman [1], Franz [5], Loetsch et al. [9]). Fondamentalement, il s'agit en effet de rechercher une série de courbes de croissance (croissance en hauteur, croissance en circonférence, croissance en volume, etc.), qui peuvent être ajustées notamment par la méthode des moindres carrés.

Il faut remarquer toutefois que la réalisation, pour les différentes caractéristiques considérées, d'*ajustements indépendants* les uns des autres doit inévitablement conduire à des discordances non négligeables (volume ajusté trop ou trop peu élevé par exemple, par comparaison avec le nombre ajusté d'arbres et avec leur circonférence et leur hauteur moyennes).

Nous voudrions mettre ici en évidence l'intérêt que peut présenter dans ce domaine la théorie des *équations de régression simultanées*, qui a été progressivement développée durant les quarante dernières années dans le secteur économique. Un travail analogue a été publié récemment par Furnival et Wilson [6] et une brève allusion aux modèles récursifs a été faite par Decourt [4].

Au cours du paragraphe 2, nous présenterons quelques modèles simples de croissance de peuplements forestiers. Au cours du paragraphe 3, nous donnerons quelques indications relatives à l'estimation des paramètres. Enfin, nous discuterons brièvement les avantages et les inconvénients de l'emploi d'équations de régression simultanées en matière d'étude de la production forestière (paragraphe 4).

2. Quelques modèles de croissance

Nous avons abordé l'emploi d'équations de régression simultanées à l'occasion de l'élaboration de tables de production provisoires relatives à l'épicéa commun (*Picea Abies* Karst.) (Dagnelie et al. [2]). Les variables étudiées dans ce cas relativement élémentaire étaient l'âge, le nombre d'arbres par hectare, la circonférence moyenne des 100 plus gros arbres par hectare (circonférence moyenne dominante), la hauteur moyenne des 100 plus gros arbres par hectare (hauteur moyenne dominante), la surface terrière par hectare (c'est-à-dire la surface totale des sections horizontales des troncs à une hauteur de référence, qui est ici 1,5 m), le volume par hectare, l'accroissement en surface terrière et l'accroissement en volume par hectare et par an. Pour les besoins du présent exposé, nous simplifierons encore le problème, en ne considérant que *cinq variables:*

x_1 = âge (à partir de la plantation),

x_2 = nombre d'arbres par hectare,

x_3 = hauteur moyenne dominante (m),

x_4 = surface terrière (m^2/ha),

x_5 = volume (m^3/ha).

Un *premier modèle théorique* peut être constitué comme suit:

$$\begin{cases} x_3 = \beta_{30} + \beta_{31}\,x_1 + d_3\,, \\ x_4 = \beta_{40} + \beta_{41}\,x_1 + \beta_{42}\,x_2 + d_4\,, \\ x_5 = \beta_{50} + \beta_{53}\,x_3 + \beta_{54}\,x_4 + d_5\,. \end{cases}$$

La première équation est une équation de croissance en hauteur, exprimée uniquement en fonction de l'âge; la deuxième une équation de croissance en surface terrière, exprimée à la fois en fonction de l'âge et de la densité du peuplement (nombre d'arbres par hectare); et la troisième une équation de cubage, faisant intervenir à la fois la hauteur et la surface terrière. Les paramètres β_{30}, β_{40} et β_{50} sont des ordonnées à l'origine, tandis que les coefficients β_{31}, β_{41}, etc. sont des coefficients de régression, simple ou multiple. Quant aux quantités d_3, d_4 et d_5, elles représentent les variables résiduelles d'écart relatives aux différentes équations. Enfin, toutes les équations sont supposées linéaires, ce qui peut être admis en première approximation, pour autant que les variables aient toutes subi une transformation logarithmique.

Un *deuxième modèle*, vraisemblablement moins adéquat au point de vue forestier, mais intéressant pour la suite de l'exposé, est:

$$\begin{cases} x_3 = \beta_{30} + \beta_{35}\, x_5 + d_3\,, \\ x_4 = \beta_{40} + \beta_{42}\, x_2 + \beta_{45}\, x_5 + d_4\,, \\ x_5 = \beta_{50} + \beta_{51}\, x_1 + \beta_{53}\, x_3 + \beta_{54}\, x_4 + d_5\,. \end{cases}$$

La première équation, mettant en relation la hauteur et le volume, serait ici une équation de qualité des peuplements; la deuxième, qui tient compte du nombre d'arbres, de la surface terrière et du volume, une équation de densité; et la troisième serait également une équation de cubage.

Moyennant une légère modification, on peut imaginer enfin la construction d'un *troisième modèle*, très semblable aux deux précédents:

$$\begin{cases} x_3 = \beta_{30} + \beta_{31}\, x_1 + \beta_{35}\, x_5 + d_3\,, \\ x_4 = \beta_{40} + \beta_{42}\, x_2 + \beta_{45}\, x_5 + d_4\,, \\ x_5 = \beta_{50} + \beta_{53}\, x_3 + \beta_{54}\, x_4 + d_5\,. \end{cases}$$

3. Quelques principes d'estimation

3.1 *Principes généraux*

Dans les modèles économiques à plusieurs équations, on introduit en général a priori une distinction entre *variables endogènes et exogènes*. Les premières sont celles dont on cherche précisément à déterminer les valeurs à l'aide des équations, tandis que les secondes peuvent être déterminées en fonction d'éléments extérieurs au modèle: cette distinction est analogue à celle qui est faite, en matière de régression simple et multiple, entre variables dépendantes et indépendantes (ou explicatives). Dans les trois exemples considérés ci-dessus, la hauteur, la surface terrière et le volume (x_3, x_4 et x_5) sont indiscutablement des variables endogènes (ou dépendantes), tandis que l'âge et le nombre d'arbres (x_1 et x_2) peuvent être considérés comme des variables exogènes (ou explicatives).

Normalement aussi, dans les modèles à équations simultanées, le *nombre d'équations* doit être égal au nombre de variables endogènes. Tel est bien le cas pour nos trois exemples.

En outre, un *symbolisme* particulier est en général adopté pour distinguer les variables exogènes et endogènes (x et y par exemple) et leurs coefficients (β et γ par exemple) et des *conditions* relativement strictes, de normalité et d'indépendance des résidus notamment, sont habituellement admises. Nous ne nous étendrons pas outre mesure sur ces points particuliers, que nous développons de façon plus détaillée dans un autre travail (Dagnelie [3]) et qui figurent en bonne place dans les ouvrages classiques d'économétrie (Johnston [7], Leser [8], Malinvaud [10], Wonnacott et Wonnacott [13], etc.).

3.2 *Les modèles récursifs*

En ce qui concerne l'estimation des paramètres, une distinction fondamentale doit être faite entre modèles récursifs et non récursifs. Les *modèles récursifs* sont tels que les variables endogènes n'interviennent que progressivement dans les différentes équations, la première équation ne comprenant qu'une variable de cette catégorie et chacune des autres équations ne faisant intervenir qu'une variable endogène supplémentaire, par rapport aux équations précédentes. Il en est ainsi pour notre premier modèle, dont la première équation ne contient que la variable endogène x_3, la seconde x_4 et la troisième, outre x_3 et x_4, la variable x_5. Il n'en est pas ainsi, par contre, pour les deux autres modèles.

Dans le cas particulier des modèles récursifs, l'estimation des paramètres peut se faire de proche en proche à l'aide des formules classiques de la régression simple et multiple. On notera toutefois que, pour assurer la cohérence de l'ensemble du système, chaque variable endogène supplémentaire doit être mise en relation avec les variables exogènes intervenant dans l'équation considérée et, le cas échéant, avec les *valeurs estimées* des autres variables endogènes qui y apparaissent, ces valeurs estimées pouvant être déduites chaque fois des équations précédentes.

C'est ainsi que, toujours pour le premier modèle présenté ci-dessus, la hauteur (x_3) doit être mise en rapport avec l'âge des peuplements (x_1), la surface terrière (x_4) en rapport avec l'âge et le nombre d'arbres (x_1 et x_2), et le volume (x_5) en rapport avec les valeurs estimées de la hauteur et de la surface terrière ($\hat{x}_3$ et $\hat{x}_4$), ces valeurs pouvant être déduites des deux premières relations. Pour les 86 parcelles observées dans l'un des deux types de forêts décrits dans notre travail initial (Dagnelie et al. [2]), on obtient, par régression simple:

$$x_3 = 0{,}297\,81 + 0{,}636\,26\, x_1\,,$$

par régression double:

$$x_4 = -\,1{,}321\,70 + 0{,}634\,77\, x_1 + 0{,}257\,16\, x_2\,,$$

et par régression double encore, après estimation des 86 valeurs estimées de x_3 et de x_4:

$$x_5 = -0{,}33375 + 1{,}07494\,\hat{x}_3 + 0{,}18465\,\hat{x}_4\ ^{1)}.$$

Si l'on avait calculé la dernière équation indépendamment des deux autres, à partir des valeurs observées de x_3 et x_4, on aurait obtenu, au contraire:

$$x_5 = -0{,}10116 + 0{,}82189\,x_3 + 0{,}97848\,x_4,$$

et on peut montrer assez simplement que cette relation n'est pas en concordance avec les deux premières équations du système (Dagnelie [3]).

3.3 *Les modèles non récursifs*

Les modèles non récursifs nécessitent l'emploi d'une méthode d'*estimation indirecte* sensiblement plus complexe. Dans chaque cas, le système doit tout d'abord être mis sous *forme réduite*, c'est-à-dire être explicité par rapport aux variables endogènes. Ceci est généralement possible, puisque le nombre d'équations est égal au nombre de variables endogènes.

Ensuite, les variables endogènes étant exprimées uniquement en fonction des variables exogènes, les différentes équations de la forme réduite peuvent être traitées indépendamment les unes des autres par régression multiple. Enfin, vient le retour au modèle initial.

Des difficultés d'*identification* des paramètres peuvent toutefois se présenter à ce stade du calcul. En ce qui concerne notre deuxième modèle par exemple, on peut montrer que le nombre de paramètres intervenant dans la troisième équation est trop élevé pour qu'ils puissent tous être estimés à partir de la forme réduite: cette équation est dite *sous-identifiée* (Dagnelie [3]). Une difficulté d'un autre ordre apparaît en ce qui concerne la première équation, qui possède un trop petit nombre de paramètres et qui est dite *suridentifiée*: cette difficulté peut toutefois être levée par l'emploi d'autres méthodes d'estimation, telle que la méthode des *doubles moindres carrés*, qui est décrite dans la plupart des ouvrages classiques d'économétrie.

Aucune difficulté particulière ne se présente par contre en ce qui concerne la deuxième équation de ce modèle, ni l'ensemble des équations du troisième modèle. Pour les mêmes données que ci-dessus, les résultats relatifs à ce troisième modèle sont d'ailleurs:

$$\begin{cases} x_3 = 0{,}12986 + 3{,}6648\,x_1 - 4{,}2343\,x_5, \\ x_4 = -1{,}11736 + 0{,}21950\,x_2 + 0{,}79222\,x_5, \\ x_5 = -0{,}18031 + 0{,}88945\,x_3 + 0{,}88104\,x_4. \end{cases}$$

1) Dans l'interprétation de ces résultats, on ne devra pas perdre de vue le fait que les données initiales ont été remplacées par leurs logarithmes décimaux.

4. Discussion et conclusions

L'emploi de la théorie des équations de régression simultanées peut paraître assez séduisant à première vue, pour assurer de façon systématique la cohérence des différents éléments figurant dans les tables de production forestières. Il ne faut pas sous-estimer cependant l'importance de certaines *difficultés* et de certaines *limitations*.

Au niveau de la construction du modèle, il est indispensable de concilier les impératifs forestiers et statistiques, pour aboutir à un système qui, à la fois, soit logique et ne présente pas de défauts insurmontables d'identification. Au niveau de l'estimation et de l'interprétation des paramètres, il y a lieu d'être particulièrement attentif aux conditions de linéarité des relations, de stabilité des variances, etc., certaines de ces conditions pouvant être très contraignantes.

Mais l'existence même de ces difficultés et de ces limitations peut augmenter l'intérêt d'éventuelles études de régression simultanée. Par l'examen des résidus notamment, l'étude comparative de différents modèles et de différentes transformations des variables initiales doit en effet permettre de contrôler la validité de certaines hypothèses de base, qui sont toujours faites, au moins implicitement, lors de la construction de tables de production.

Bibliographie

[1] Assman, E. (1961): Waldertragskunde. BLV, Munich, 490 p.
[2] Dagnelie, P., Nivelle, J. L., Rondeux, J. et Thill, A. (1970): Production de l'épicéa commun (*Picea Abies* Karst.) dans quelques stations de l'Ardenne Centrale. Bull. Rech. Agron. Gembloux *5*, 428–442.
[3] Dagnelie, P. (1975): Analyse statistique à plusieurs variables. Presses Agron., Gembloux, 362 p.
[4] Decourt, N. (1972): Méthode utilisée pour la construction rapide de tables de production provisoires en France. Ann. Sci. Forest. *29*, 35–48.
[5] Franz, F. (1966): Zum Aufbau neuzeitlicher Ertragstafeln. Forstwiss. Centralbl. *85*, 129–147.
[6] Furnival, G. M. et Wilson, R. W. (1971): Systems of equations for predicting forest growth and yield. In: Patil, G. P., Pielou, E. C. et Waters, W. E.: Statistical ecology (3 vol.). Pennsylvania State Univ. Press. *3*, 43–57.
[7] Johnston, J. (1963): Econometric methods. McGraw Hill, New York, 300 p.
[8] Leser, C. E. V. (1966): Econometric techniques and problems. Griffin, Londres, 119 p.
[9] Loetsch, F., Zöhrer, F. et Haller, K. E. (1973): Forest inventory (2 vol.). BLV, Munich, 436 p. et 469 p.
[10] Malinvaud, E. (1969): Méthodes statistiques de l'économétrie. Dunod, Paris, 782 p.
[11] Schwappach, A. (1912, 1923 et 1929): Ertragstafeln der wichtigeren Holzarten in tabellarischer und graphischer Form. Neumann, Neudamm, 74 p.
[12] Wiedemann, E. (1938, 1949 et 1957): Ertragstafeln wichtiger Holzarten bei verschiedener Durchforstung. Schaper, Hanovre, 194 p.
[13] Wonnacott, R. J. et Wonnacott, T. H. (1970): Econometrics. Wiley, New York, 445 p.

Adresse de l'auteur:
Pierre Dagnelie, Faculté des Sciences Agronomiques de l'Etat, Gembloux, Belgique.

H. Riedwyl and U. Kreuter

H. Riedwyl

Identification

Summary

The identification is a method of assigning a new member to a population on the basis of a set of variables observed. It is shown that the identification analysis can be taken over from a multiple regression approach which is already well known for the discrimination of two groups.

1. Introduction

We are given a sample from a population. The measurements of a certain number p of variables are determined for each sample member. A method of assigning a new member to the population on the basis of the measurements of the p variables we call an identification method. If we have more than one population in order to assign a new member, this problem is often referred to discrimination. In the simplest case with two populations, following Fisher [2], we construct a linear discriminant function of the variables $x_1, x_2, \ldots, x_p$

$$X = \sum_{j=1}^{p} \beta_j x_j . \tag{1}$$

The coefficients β_j are derived on the assumption that the two populations with n_1 and n_2 observations each are multivariate normal with identical dispersion matrices. Fisher introduced a pseudo-dependent variable y_i $(i = 1, 2, \ldots, n_1 + n_2)$

$$y_i = \begin{cases} \dfrac{n_2}{n_1 + n_2} & \text{for all members of the first group} \\[2ex] -\dfrac{n_1}{n_1 + n_2} & \text{for all members of the second group} \end{cases}$$

to estimate from a multiple regression of y on $x_1, x_2, \ldots, x_p$ the coefficients β_j which would in fact only differ from those written above (1) by a constant factor.

2.

In the case of a multivariate normal population, denoting by the p-vector variable X with mean μ and variance-covariance matrix Σ, let $\bar{\boldsymbol{x}}$ be the sample mean vector, S the dispersion matrix, $\boldsymbol{x}$ the vector of the new member, $\boldsymbol{d} = \boldsymbol{x} - \bar{\boldsymbol{x}}$, then

$$\chi^2 = \boldsymbol{d}' S^{-1} \boldsymbol{d} \tag{2}$$

is asymptotically chisquare distributed with p degrees of freedom. For the regression approach to identification we can calculate χ^2 from the coefficient of determination R^2 by

$$\chi^2 = \frac{n^2 - 1}{n} \frac{R^2}{1 - R^2} . \tag{3}$$

Terming it according to Bartlett [1], the multiple regression approach to discrimination problems is not only possible but also has the advantage that all the theory known in regression can be taken over to the discrimination case. In this way, Bartlett considered the analysis of variance of pseudo-variable and discussed standard errors of the discriminant function coefficients.

A method of selection of the best set of variables to predict the dependent variable in multiple regression can be interpreted in discrimination as looking for a set of variables which best discriminate the two populations. If you have to assign a new member to one of the two populations, the discrimination approach allocates it to either one of them. The theory does not divide the sample space of the discrimination function into three regions, two for the populations and one for an area of indecision.

In the following part, we consider identification as a special problem in discrimination where one group has only one member. If the group has n members, we construct the pseudo-variable y with $n_1 = n$ and $n_2 = 1$. In constructing such a procedure of identification, we again have the advantage that all the theory of regression can be taken over to the identification case.

3. An example

We measured six physical characteristics on a group of 100 banknotes:

x_1: length of the banknote,
x_2: width on the left side,
x_3: width on the right side,
x_4: width of the lower margin,
x_5: width of the upper margin,
x_6: diagonal length of the interior print.

The dispersion matrix S of the six variables is given in Table 1.

Table 1
Dispersion matrix S of 100 banknotes.

x_1	x_2	x_3	x_4	x_5	x_6
15.0241	5.8013	5.7293	5.7126	1.4453	0.5482
	13.2577	8.5899	5.6652	4.9067	− 4.3062
		12.6263	5.8182	3.0646	− 2.3778
			41.3207	− 26.3475	− 0.0187
				42.1188	− 7.5309
					19.9809

3.1 Five new banknotes which were probably false, we identify by discriminating each one from the population. Table 2 gives the chisquare values (3) indicating in brackets the importance of the most significant variables by a forward selection procedure.

Table 2
Chisquare values of the identification analysis for the five banknotes and a true outlying banknote.

1	2	3	4	5	true banknote
14.8 (x_6)	18.4 (x_6)	14.8 (x_6)	36.9 (x_6)	48.6 (x_6)	17.1 (x_5)
22.5 (x_4)	32.3 (x_4)	18.3 (x_5)	38.9	48.8	18.9
39.1 (x_5)	54.5 (x_5)	19.7	39.5	49.1	19.8
45.1 (x_2)	64.5 (x_2)	20.8	39.8	49.6	20.2
45.4	66.7	21.2	39.8	49.9	20.3
45.5	66.8	21.2	39.8	49.9	20.3

From the fact that x_6 is always present after eliminating the redundant variables, it seems evident that the five banknotes are falsifications by the same falsifier.

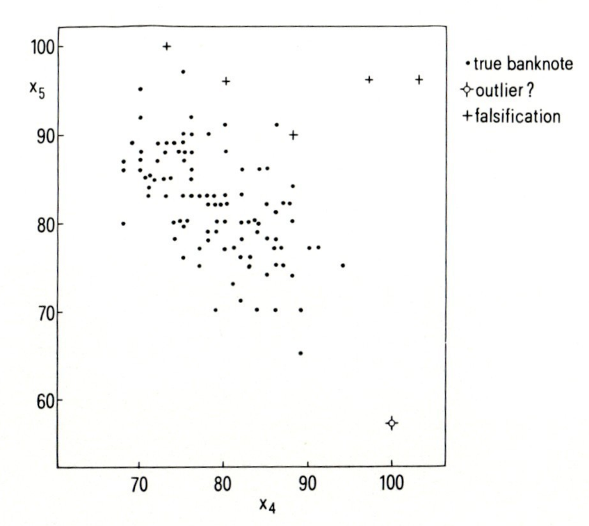

Fig. 1

3.2 Identification analysis is also useful in detection of an outlier in multivariate analysis. The type of multivariate outliers may be faulty because of a gross error in one of its variables or because of systematic error in several variables. In Fig. 1 a true banknote could be an outlier. The analysis of identification as in 3.1 results in the fact that this banknote is not typical of the population because of variable x_5 (see Table 2).

References

[1] BARTLETT, M. S. (1965): Multivariate statistics in theoretical and mathematical biology (ed. T. H. Waterman and H. J. Morowitz). Blaisdell Publishing Company, New York.

[2] FISHER, R. A. (1936): The use of multiple measurements in taxonomic problems. Ann. Eug. 7.

Address of the authors:
Institut für mathematische Statistik und Versicherungslehre der Universität Bern, Sidlerstrasse 5, CH-3012 Bern.

Francesco Brambilla

Francesco Brambilla

Stability of Distance Between Structured Groups in a Social Organism: Empirical Research[1])

Summary

The working hypothesis at the basis of the research is that in the development process of a social body the distance between structural groups of persons remain constant. The structural groups considered are men, women and a selected group of women (Soroptimists). The inquiry was carried out in 15 European countries in 1972. The total number of interviews was 4200. The six variables considered in the inquiry are the attitudes in respect of work, family, education, free time, sex and politics. The discriminatory analysis techniques are: entropy and factor analysis. Results seem to confirm the hypothesis of stability between groups and of countries between one another.

1. A nation as a society is a group of people bound by a common set of laws, institutions, way of life, traditions, common language, common means of communications, and so forth.

The individual becomes conscious of this complex structure not through scientific analysis but through an awareness of signals (latu sensu), which is subject to a metabolism of decisions, attitudes, or actions relative to the outside world. Consequently, decisions are thus the transformer which guarantees the growth of a society through the metabolism constituted by a signal (variation of the state of the outside world).

2. This growth takes two basic directions: development and expansion. Development can be expressed in terms of variations of the frequency and the types of mutual relations between individuals (physical and juridical persons), while the expression and realization of expansion lies in the activation and the interlinking of these relations.

[1]) All figures and other information in the 'Responsability of Women in Contemporary Society'. A copy of this complete report can be supplied on request from Mrs. Lida Brambilla-Longoni, Via Erica 49, Pineta di Arenzano, Arenzano, Genova (Italy).

3. One of the problems which thus far has not been sufficiently examined, and which is the subject of this investigation, is the one inherent in the *distance* between the groups of persons which constitute the elements in the formation of a national entity.

4. Our investigation—which is empirical because based essentially on statistical analysis of observation of elements gathered for this purpose—purports to prove a working hypothesis, namely, that in the process of its information, society leaves unchanged the distances between structural groups of individuals.

The structural groups—that is to say those linked through a common denominator of norms, attitudes, and so forth—which our investigation sets out to examine, are men and women in general, as well as an association of women with identical associative and selective structure in all European countries in which research has been conducted. This is the Soroptimist International, which in 1972 undertook an investigation within the European Federation of Soroptimists Clubs, collected the data obtained thanks to the initiative of the then President of the Federation, Lida Brambilla, and organized the material.

This writer, on the other hand, was in charge of the statistical analysis, with the cooperation of the Institute of Quantitative Methods of Bocconi University.

5. The technique used to gather information was to ask each interviewee, in strictly random sequence, ten questions structured in six scales of *attitudes* formulated identically for all member countries of the European Federation of Soroptimist International, and designed not to elicit information on concrete problems through what may be called the conscious method, but to encourage the self-classification of each interviewee with respect to her acceptance of the system of values in which she lives.

The questions asked required a choice between a positive or a negative answer to six fundamental aspects and problems of daily life, to wit:

$X1$ – Women and work (scale A)
$X2$ – Women and family (scale B)
$X3$ – Women and education and upbringing (scale C)
$X4$ – Women and free time (scale D)
$X5$ – Women and sex (scale E)
$X6$ – Women and politics (scale F).

The attitude is the identifiable expression of that *globality* with which we live through our daily experiences.

The scales of attitudes represent the technical means of *bringing* this globality to light.

They are *not* a collection of *opinions, nor* do they represent a measure of anything.

Through the number of positive answers given by each interviewee to each scale of values, he or she *unconsciously* places himself in one of the groupings ranging from greatest to least resistance to changes in the organization of the society in which he or she lives, seen in relation to the six recorded trends.

It is precisely this globality come to light which makes it possible to compare different social positions, different sexes, different social and economic structures.

Furthermore, many of the questions, which appear to be put to the interviewees in order to elicit information on individual aspects of the problem and therefore to provoke perplexity, must be seen in this light.

In order to clarify plastically how the answer to each question—which is an oral attitude—is the unconscious total expression of a complex circuit of relations with the society in which the interviewee lives, we have presented in the Figures 1 and 2 the structure of an agricultural and of an industrial culture, respectively, in the sociological sense.

In the circuit *A*, the initial emphasis rests upon the relationship woman-home; in the second, the initial point of emphasis is the modern woman and her various interests outside her home. In the first, strong ties with the parents and a strong social control prevail, while in the second we find a minor social control. In the first, outside elements connected with a choice are of the utmost importance; the second is marked by a greater sense of adventure. These elements which always characterize circuit *A* present *two* trends of development:

5.1 The first is manifest in the relationship with *mother* and family in an *agricultural society*; in the industrial society, on the other hand, it is expressed in the relationship with the *husband* (understood not as a continuation of the father figure, but as a friend), as well as with *colleagues* and *friends*.

5.2 The second becomes manifest in the way in which the interviewee judges an out-of-the-ordinary situation (divorce, permissiveness in manners and customs, etc.), which is binding in one society and does not represent a stumbling block in the other.

There is a second circuit in the formation of attitudes (circuit *B*) which, activated by the reading of magazines, reviews and books, derives from people's different levels of insecurity vis-a-vis emancipation.

There is a strong degree of insecurity in an agricultural culture aspiring to a freer life and consequently strongly influenced by the way of life of an industrial society as soon as they come in contact[1]).

There is less fear of not belonging to the 'establishment' and awareness of living in a lively and valid culture, in an industrial society.

A third circuit in the formation of attitudes (circuit *C*) is connected with the influence of so-called hidden persuaders, centered in an agricultural society in the person of the *priest*, and in an industrial society in that of the *psycho-analyst*.

1) In this respect, a high territorial mobility and circulation of consumer's goods, consequent to the great mass media increase, have sped up the comparison process between the two cultures.

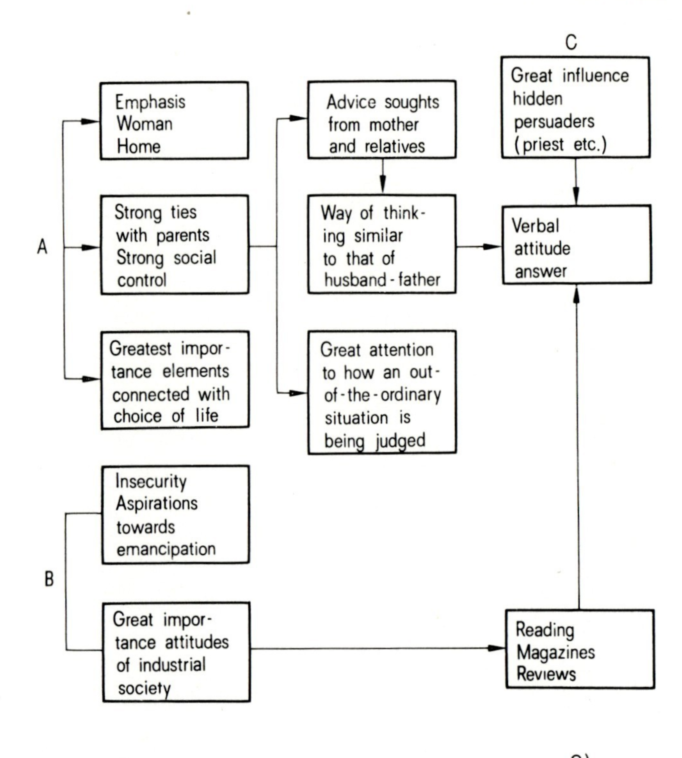

Fig. 1
Agricultural society.

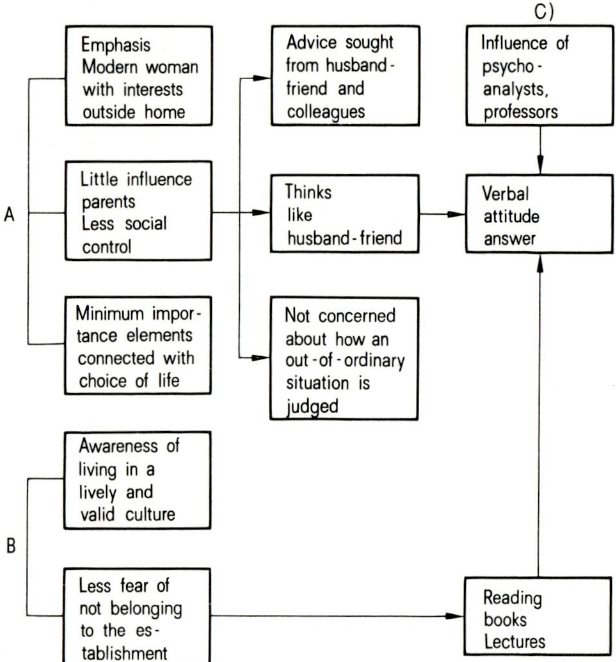

Fig. 2
Industrial society.

The scales of attitudes in their different formulations through 58 questions tented to relate to the various circuits as if they had been prompted by different sets of rules. If all interviewees answer *no* to the various questions, the score is zero, which means there is a maximum degree of *resistance* to emancipation. The increase in the positive score means *growing acceptance of emancipation.*

6. Hierarchy within the scales

Emancipation has been examined through six different trends constituted by the six scales of values. It is to be expected that the tendency towards emancipation will *become evident* in different ways, depending on the formulation of the questions, and that, above all, emancipation will be accepted with different degrees of enthusiasm, depending on the *spheres of interests involved.*

The results of the investigation are highly interesting, because they show up a hierarchy of resistance in Soroptimists as well as in non-soroptimists women and in men, a hierarchy which is indicated in the following scales ranked according to the increasing degree of resistance to emancipation:

- *Politics and education and upbringing* (scale F and C)
- *Family* (scale B)
- *Work* (scale A)
- *Free time* (scale D)
- *Sex* (scale E).

This order of rank has been shown to be valid for all countries and sexes.

For each country and sex, that is to say for each line (0 = Soroptimist; 1 = non-Soroptimist women; 2 = men) and for each scale (which must be read by *column*) two numbers are given: the first relates to the absolute number of positive answers, the second to the relative percentage.

A synthetic measure of the six scales, known in statistics as 'entropy', to which we will come back later.

A *reclassification* of the scales—always for each country and sex—according to the decreasing value of the percentages, gives us Table 1, where we note that the scales rank from that of major percentage (i.e. less resistance to emancipation) to that of minor percentage (i.e. greater resistance).

Examining the date we note that the rank order of scales is exactly the one indicated above.

7. Differentiation between countries and sexes

Even taken for granted the interviewee's greater or lesser resistance according to the private sphere of interests to which the various scales relate, each interviewee has answered *globally* with respect to his or her *individual position* vis-a-vis the problem of *feminine emancipation.* It is therefore necessary to *synthetize* appropriately the *six* results (one for each scale), which express this '*level of emancipation*'.

Obviously, the synthesis can be different. Those *examined thus far* relate to two different branches of statistical methodology. They are presented in appropriate appendixes:

1. Entropy analysis.
2. Factor analysis.

Differentiation in terms of entropy

The values of entropy defined as $H = -\sum_{i=1}^{6} p_i \log p_i$ (with p_i as percentage of positive answers to scale i) listed in Table 1 show up clearly not only the difference *between countries*, but also that *between sexes* in each country. To help understand the results, we present them in graphic form (Graph no. 1).

Each country has a value of entropy for each sex. If we mark this value with a point on a line connecting minor and major values, we obtain for each *sex* a *differentiation* between countries as indicated in the graph.

Interpretation of entropy

To make it understandable why we have used entropy as information unit contained in the answers, we are presenting here an interpretation in terms of our problem.

Each questions listed in the questionnaire can be taken as a message made up of N *stimuli*. If there are P stimuli of category A available (i.e. P stimuli towards the awareness of the system of values of the society in which the interviewee lives) and $N - P = Q$ stimuli of category B (i.e. Q stimuli *incapable* of directing the interviewee towards that awareness), the number of different messages which can be obtained through this list of symbols is given by the *number of combinations* that can be made with these N stimuli by taking them each as groupings formed each by P stimuli. This number of combinations is shown with the mathematical symbol of $\binom{N}{P}$ stimuli.

It thus becomes clear that entropy H is simply *equal to the quantity* $\log \binom{N}{P} / N$ i.e. it is proportional to the logarithm of these combinations.

Differentiation in terms of factor analysis

A second approach to the problem of differentiation has been that of latent structure. The methods of Hotelling's principal components was employed.

This analysis has made it possible to classify each country (meaning for each sex, age, etc.) in terms of two factors F_1 and F_2, which have the following significance:

1. F_1 = grade of *resistance* to *emancipation*.
2. F_2 = grade of *propensity* towards *emancipation*.

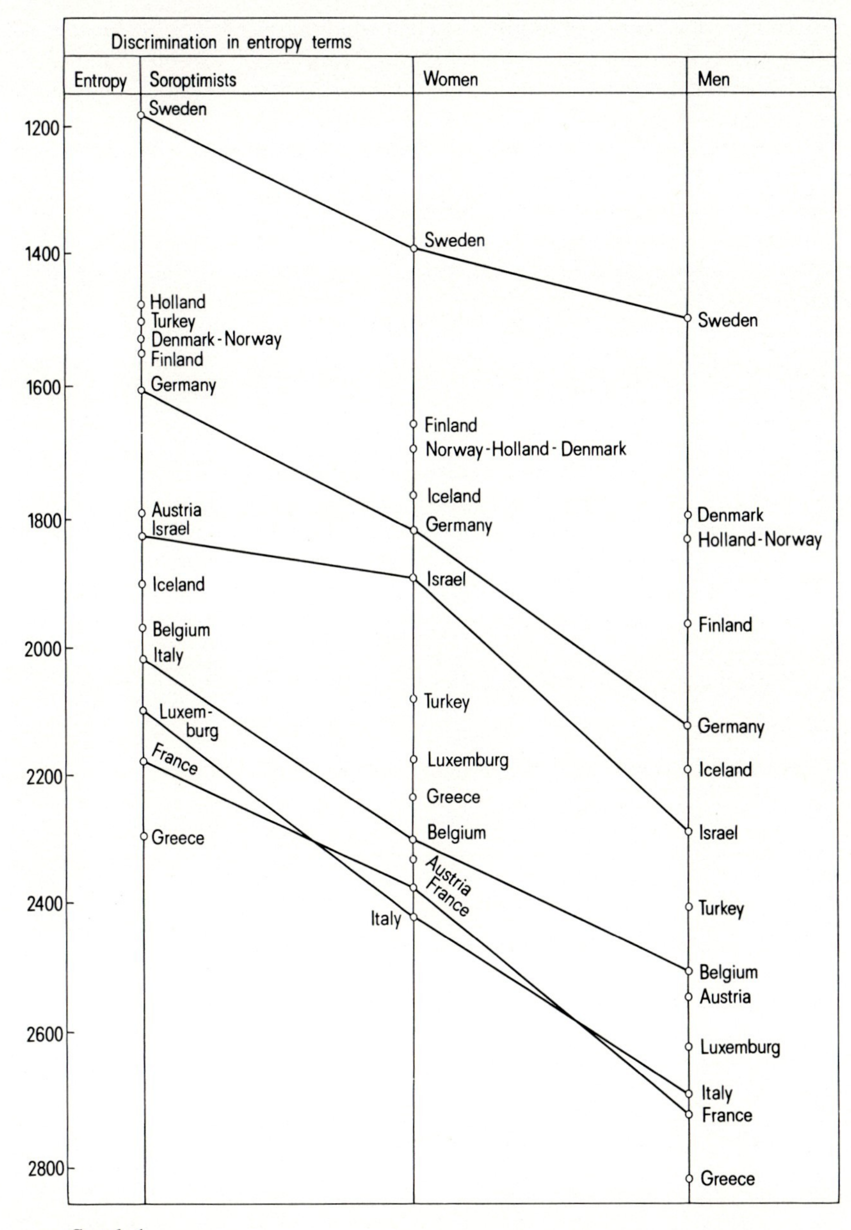

Graph 1

The two factors are not one the repetition of the other. The first relates to the *global attitude toward the innovation of the entire system* of values forming the society in which one lives; the second factor, on the other hand, is the expression of the *degree of will to act* toward an innovating direction.

In our case, this will express itself in making explicit a *propension* toward women's emancipation, both by stating it by a clear-cut opinion, or by a political action (for instance, by actively taking part to the life of an association such as Soroptimism).

Both F_1 and F_2 have the logical dimensions of an energy and their sum $F_1 + F_2 = C$ can there be thought of as an index of the complexive degree of tendency to emancipation (indicated by C).

In view of the fact that the relation $F_1 + F_2 = C$ makes a straight line, all points representative of the countries having the same C lie on the same straight line.

Each country is therefore identified by the two values, as shown in Table 1. Each *country* is further identifiable with a *point* on the line between F_1 and F_2.

The main results drawn from the above analysis are:

1. Soroptimist women form a group sharply different from the other two in their favourable attitude toward women's emancipation.
2. Non-Soroptimist women are placed between Soroptimists and men.
3. Men show greater resistance to women's emancipation, as compared to women.
4. Within each group, the distances between countries are the same.

Table 1
Discrimination of Countries in Terms of Factor Analysis.

Countries	Soroptimists		Women		Men	
	F_1	F_2	F_1	F_2	F_1	F_2
Austria	3.00	1.77	3.25	1.57	3.99	1.21
Belgium	2.99	1.59	3.37	1.48	3.09	1.43
Denmark	2.65	1.62	3.61	1.25	3.26	1.53
Finland	2.64	1.62	2.56	1.74	3.47	1.25
France	3.25	1.39	3.23	1.42	3.58	1.24
Germany	2.20	2.14	3.20	1.57	3.82	1.20
Greece	2.89	1.92	3.01	1.72	3.67	1.24
Holland	2.65	1.66	3.01	1.45	3.14	1.40
Iceland	4.18	1.51	2.73	2.20	3.83	1.58
Israel	2.38	1.87	3.53	1.37	3.39	1.53
Italy	3.44	1.26	3.95	1.00	4.34	0.82
Luxemburg	2.65	2.33	2.69	2.00	4.05	1.35
Norway	2.90	1.58	2.66	1.66	3.51	1.26
Sweden	1.78	2.07	2.51	1.80	2.99	1.52
Turkey	2.98	2.14	3.95	1.29	4.13	1.13

Coordinate F_1 = Degree of resistance to emancipation
Coordinate F_2 = Degree of propension to emancipation

Author's address:
Professore Francesco Brambilla, Università Commerciale Luigi Bocconi, Istituto Metodi Quantitativi, Via R. Sarfatti 25, Milano (sud 13/24), Italia.

6. Miscellaneous

I. M. Chakravarti

I. M. Chakravarti

Statistical Designs from Room's Squares with Applications

Summary

Combinatorial properties of various experimental designs derived from Room's squares, relevant to statistical analysis of data from such designs have been examined. It is noticed in particular that a balanced Room's square provides us with a doubly balanced design which can be used to estimate residual effects of treatments in a sequence of experiments using the same experimental units. It is shown how incomplete paired-comparison designs with a certain degree of symmetry can be derived from Room's squares. Different approaches to analysis of data from paired-comparison designs for the purpose of establishing an order or partial order among the objects compared are examined. In particular it is noted that 'Multidimensional Scaling techniques' can be used on the dissimilarities between pairs of objects ascertained from paired-comparison designs to establish an order or partial order between the objects.

1. **Introduction**

A *Room's square* design [23] of order $2n$, n an integer, is an arrangement of $2n$ elements in a square array of side $2n - 1$ such that

a) each of the $(2n - 1)^2$ cells of the array is either empty or contains an unordered pair of distinct elements,
b) each of the $2n$ objects appears precisely once in each row and in each column, and
c) each unordered pair of distinct elements occurs in exactly one cell of the array.

A balanced incomplete block design with parameters (v, b, r, k, λ) is an arrangement of v varieties in b blocks of k distinct varieties such that each

This research was partly supported by an NSF grant GP–42325.

variety is contained in r blocks and each pair of varieties is contained in λ blocks.

If in a Room's square every unordered pair is replaced by an ordered pair and from every line of this *ordered Room's square*, two blocks are formed—one consisting of the first elements of the ordered pairs and the other consisting of the second elements, we get an incomplete block design in $2(2n-1)$ blocks which is self complementary. If the self complementary incomplete block design is a balanced incomplete block design then the corresponding ordered Room's square is called a *balanced Room's square* [19]. A doubly balanced or a 3-design is an arrangement of v elements into blocks of k distinct elements such that every unordered triple of distinct elements occurs in precisely μ blocks. It has been shown [19] that every self complementary balanced incomplete block design is a doubly balanced or a 3-design. Hence a balanced Room's square gives rise to a 3-design.

In a duplicate bridge tournament for $2n$ teams, a complete Howell rotation is an arrangement such that

(i) each board is played by at most one pair of teams on any round
(ii) every team plays one board every round and every team plays each board precisely once and
(iii) every team opposes every team precisely once.

It is clear that a Room's square of order $2n$ is equivalent to a Howell rotation for $2n$ teams. Suppose now that every board has a NS direction and EW direction. Team 1 is said to compete against Team 2 on a particular board if they play the board in the same direction.

A complete Howell rotation for $2n$ teams is called a complete balanced Howell rotation if the arrangement also satisfies the condition

(iv) each team competes equally often, say λ times, with every other team.

It has been shown [19] that a balanced Room's square is equivalent to a complete balanced Howell rotation. Balanced Howell rotations for $4t = p^r + 1$ teams where $p^r > 3$ is a prime power have been constructed [4]. Other constructions are given in [19].

Our interest in these combinatorial configurations is in their possible use as experimental designs in calibrations or comparisons for ranking using a suitable statistical model. These designs will be useful in biological and medical research for evaluation of techniques and drugs using matched experimental units and also in situations where a sequence of experiments has to be performed on the same set of experimental units keeping in mind that residual treatment effects can not be ignored.

2. Room's Squares as Experimental Designs

Consider the following Room's square of order $2n = 8$ taken from [1] after renumbering the columns so that there is no entry in the principal diagonal.

Table 1
A Room's square of order 8.

	1	2	3	4	5	6	7
1	–		37	56	01		24
2	35	–	02	41		67	
3		46	–	03		52	71
4	12			–	57	04	63
5	74	23	61		–		05
6	06	15		72	34	–	
7		07	45		26	13	–

In [1], it has been described how the Room's square above can be used as three different types of experimental designs.

2.1 If the columns are regarded as blocks, the rows as a set of treatments $\{a_1, \ldots, a_7\}$ and the numbers in the squares as a second set of treatments $\{b_0, \ldots, b_7\}$ then the design is a balanced incomplete block ($v = 7$, $k = 4$, $\lambda = 2$) with respect to the first set of treatments and a complete randomized block with respect to the second set of treatments. Further, the design is also balanced with respect to combinations of different levels of treatment a with different levels of treatment b.

2.2 The design may be regarded as a balanced incomplete block for treatments $\{a_1, \ldots, a_7\}$ with main plots split for treatments $\{b_0, \ldots, b_7\}$.

2.3 The design may also be regarded as a supplemented incomplete block design with respect to treatments $\{a_1, \ldots, a_7\}$, as in successive experiments. Appropriate analysis of variance for each type of design is also described in [1].

The parameters of a balanced incomplete block design corresponding to a *balanced* Room's square of order $2n$ are $v = 2n$, $b = 2(2n - 1)$, $r = 2n - 1$, $k = n$ and $\lambda = n - 1$. Since this is a doubly balanced design as remarked in [19], it can be used to estimate residual effects of treatments when the same experimental units are used for a sequence of experiments [26]. In [9] it is explained how doubly balanced designs can be used for estimation of treatment effects when these are correlated. Again, a doubly balanced design gives rise to a partially balanced array [10] which is used in factorial experiments.

Consider the Room's square of Table 1. From row 1 we can form two blocks (3, 5, 0, 2) and (7, 6, 1, 4). Thus we shall get 14 blocks. It can be verified that this is a self complementary balanced incomplete block design with parameters $v = 8$, $b = 14$, $r = 7$, $k = 4$, $\lambda = 3$ and hence also a doubly balanced design in which every triple occurs exactly once.

Using the method of construction described in [1], it can be shown that we can construct such a design for every $n = 2^m$, m an integer. Complete balanced Howell rotations or equivalently balanced Room's squares of order $v = 2n = 4t = p^r + 1$, $p^r > 3$ a prime power are also available [4]. These will provide us with doubly balanced designs with parameters $v = 4t = p^r + 1$, $k = n = 2t$, in which every triple occurs $(n - 2)/2 = t - 1$ times.

3 Room's Squares as Incomplete Paired-Comparisons Designs

In a paired-comparison experiment some or all of the $\binom{v}{2} = v(v-1)/2$ pairs of v objects are presented to one or several judges. For every pair of items compared each judge may simple place the two items in the relationship preferred: not-preferred or in general may express the result of the comparison by a real number x_{ij} for the pair (i, j) compared. An extensive literature exists in this subject (see for instance [11], [18]). The simplest design for this problem most frequently treated in the literature is called the round-robin tournament in which each of v players $A_1, A_2, \ldots, A_v$ meets every other player once and each game results in a win for one of the players who receives 1 point and the loser scores 0. Very detailed references on the combinatorial as well as statistical aspects of tournaments and their results are available in [11], [12], [18].

Kendall [15] and Bose [2] constructed certain paired-comparison designs which are symmetrical with respect to objects and judges but do not require each judge to compare all possible pairs of objects. In [15], Kendall suggested certain elementary principles to be followed in the choice of a subset of $v(v-1)/2$ pairwise comparisons by a single observer:

a) every object should appear equally often,
b) the pairwise comparisons should not be divisible in the sense that we can split the objects into two sets and no comparison is made between any object in one and any object in the other.

Bose [2] defined a class of designs called linked paired-comparison designs for comparing v objects employing t judges in the following manner.

(i) Among the r pairs (all distinct) compared by each judge, every object occurs equally often say α times.
(ii) Each pair is compared by k judges $k > 1$.
(iii) Given any two judges, there are exactly λ pairs which are compared by both judges.

If each judge is considered as a treatment and each pair of objects as a block, it follows that the existence of a linked paired-comparison design of the type defined will imply the existence of a balanced incomplete block design with t treatments, $v/2(v-1)$ blocks such that each block contains k treatments, each treatment occurs in r blocks and every pair of treatments occur together in λ blocks. It then follows that

$$r = \frac{1}{2} v \alpha, \quad \lambda(t-1) = r(k-1), \quad t \alpha = k(v-1),$$

$$v \alpha \geq 2k, \quad \lambda \leq \frac{1}{2} \alpha(\alpha+1) \leq r.$$

Let

$$n_{iju} = \begin{cases} 1 \text{ if objects } i \text{ and } j \text{ are compared by judge } u \\ 0 \text{ if objects } i \text{ and } j \text{ are not compared by judge } u, \end{cases}$$

$$i \neq j = 1, 2, \ldots, v, \; u = 1, 2, \ldots, t$$

$$n_{iiu} = 0, \qquad i = 1, 2, \ldots, v, \; u = 1, 2, \ldots, t\,.$$

Let

$$N_u = (n_{iju})$$

be the incidence matrix for judge u. Then it has been shown [25] that the conditions

a) $N_u\, J = \alpha\, J \quad u = 1, \ldots, t\,,$

b) $\sum_{u=0}^{t} N_u = k\, J\,,$

c) $\operatorname{tr} N_u\, N_{u'} = 2\,\lambda \quad u, u' = 1, \ldots, v,\; u \neq u',$

where J is a matrix of order v, each of whose elements is unity and $N_0 = k\, I$, are necessary and sufficient for the existence of a linked paired-comparison design.

For the Room's square of order 8 given in Table 1, suppose we let the rows and columns correspond to seven judges ($t = 7$) and the entries $0, 1, \ldots, 7$ correspond to 8 objects. Then it is clear that every judge compares $r = 8$ pairs, every object occurs $\alpha = 2$ times among the 8 pairs compared by a judge and every pair of objects is compared by two distinct judges.

Given any two judges, the number of pairs of objects judged by both is either one or two for this design. Hence this is not a linked paired design. However, we can define a relationship between the judges in the following manner. Two judges are *first associates* if the number of pairs judged by both is 2, they are second associates if this number is 1. The following table shows the first and second associates for each judge.

Table 2

Judges	First Associates	Second Associates
1	4, 5	2, 3, 6, 7
2	3, 6	1, 4, 5, 7
3	2, 7	1, 4, 5, 6
4	1, 6	2, 3, 5, 7
5	1, 7	2, 3, 4, 6
6	2, 4	1, 3, 5, 7
7	3, 5	1, 2, 4, 6

But it can be verified that this is not a two-class association scheme as defined in [3], since constancy of p_{11}^2 is not satisfied although $p_{11}^1 = 0$ for every pair of first associates. There is no doubt that Room's squares have very interesting combinatorial properties, when viewed as an incomplete paired-comparison design for several judges.

4. Statistical Analysis of Incomplete Paired-Comparison Designs Derived from Room's Squares

In a simple paired-comparison experiment in which every object is compared once with every other object with the results expressed as a preference for one or the other object, one is interested in ranking the objects on the basis of the results of the paired comparison design. Many methods have been suggested and discussed in the literature [11], [12], [18].

A method based on the maximum likelihood principle [27], [5], assumes that each object has a positive 'worth' w_i such that in a comparison between the objects A_i and A_j the probability that A_i will be preferred to A_j is given by $w_i/(w_i + w_j)$. If we assume that the outcomes of the various comparisons are independent then the probability that a particular paired-comparison experiment E_n gives the results $\{a_{ij}\}$ where $a_{ij} = 1$ or 0 according as A_i is or is not preferred to A_j is given by

$$\Pr(E_n : w_1, w_2, \ldots, w_n) = \prod_{i<j} \left(\frac{w_i}{w_i + w_j}\right)^{a_{ij}} \left(\frac{w_j}{w_i + w_j}\right)^{a_{ji}}$$

$$= \frac{\prod_i w_i^{s_i}}{\prod_{i<j} (w_i + w_j)},$$

s_i is the number of comparisons in which A_i is preferred. The problem is to determine the values of the w_i's that maximize the probability given the results of the comparisons. If the tournament E_n is reducible, there exists no positive solution of $(w_1, w_2, \ldots, w_n)$. If E_n is irreducible then it can be shown that there exists a unique set of positive worths $w_1, w_2, \ldots, w_n$ that maximize $\Pr(E_n : w_1, \ldots, w_n)$.

Another method of ranking the objects in a complete paired-comparison design is due to Wei and Kendall [15]. Let $A = (a_{ij})$ denote the preference matrix of a paired-comparison design, $a_{ij} = 1$ or 0 according as A_i is or is not preferred to A_j, $a_{ij} + a_{ji} = 1$, $a_{ii} = 0$, $i, j = 1, 2, \ldots, n$. It is known that if the matrix A is irreducible, it is primitive for $n \geq 4$ [18]. Now if A is primitive, it follows from Frobenius' theorem [24] that

$$\lim_{i \to \infty} \left(\frac{A}{\lambda}\right)^i \boldsymbol{e} = \boldsymbol{y},$$

where λ is the unique positive characteristic root of A with the largest absolute value, $\boldsymbol{e}$ is a column vector of 1's and $\boldsymbol{y}$ is a positive characteristic vector of A corresponding to λ. Therefore, the normalized vector $\boldsymbol{y}$ can be taken as the vector of relative 'worths' of the objects in E_n if E_n is irreducible and $n \geq 4$.

Additional material on the problem of ranking a collection of objects on the basis of binary comparisons from complete paired-comparison designs may be found, for example in [6], [7], [8], [12], [13], [14], [18], [21], [22].

A method of statistical analysis of paired-comparison experiments where a 7- or 9-point preference scale is used and allowance is made for possible effects due to the order of presentation within a pair has been developed by Scheffé [20].

An analysis of incomplete paired-comparison designs of Kendall [15] and Bose [2], using Zermolo-Bradley-Terry type model has been given in [25]. It is assumed that there exist numbers $\pi_{1u}, \ldots, \pi_{vu}$ corresponding to judge $u (u = 1, 2, \ldots, t)$, where

$$\pi_{iu} \geq 0 , \quad i = 1, \ldots, v, u = 1, \ldots, t ,$$

$$\sum_i \pi_{iu} = 1, u = 1, \ldots, t ,$$

such that

$$\operatorname{Prob}(A_i \rightarrow A_j \mid u) = \frac{\pi_{iu}}{\pi_{iu} + \pi_{ju}} ,$$

$$\operatorname{Prob}(A_j \rightarrow A_i \mid u) = \frac{\pi_{ju}}{\pi_{iu} + \pi_{ju}} ,$$

where $(A_i \rightarrow A_j \mid u)$ denotes that A_i is preferred to A_j by judge u. Likelihood ratio tests, various estimation problems and large sample distributions are also considered in [25].

The method of analysis described in [25] is applicable also to the incomplete paired-comparison design derived from the Room's square.

When several judges make paired comparisons on the same set of objects, Kendall [15] suggested that we may add the preference matrices for the judges if the judges are equally reliable and apply Kendall-Wei technique to find the normalized vector $\boldsymbol{y}$ which will give us the relative 'worths' of the objects. A variant of this method will be to consider a convex mixture of the preference matrices with the weight for each matrix proportional to the degree of reliability of the judge.

Yet another technique which can be used here is the one known as 'multidimensional' scaling [16], [17], which consists in finding v points in t dimensional Euclidean space, whose interpoint distances match in some sense the experimental dissimilarities of v objects. From the paired-comparison design, suppose we have obtained a measure of dissimilarity between objects A_i and A_j as s_{ij}, $i \neq j = 1, 2, \ldots, v$. Let each of v points in t dimensional Euclidean space be associated with one of the original objects. Let the Euclidean distance between the i-th and the j-th points be denoted by d_{ij}. Let X denote the set of all pairs $x = (i, j)$, $1 \leq i < j \leq n$. Let us define a simple order $\lesssim$ on X by the requirement $(i_1, j_1) \lesssim (i_2, j_2)$ if $s_{i_1j_1} < s_{i_2j_2}$, where s_{ij} is the dissimilarity of the pair (i, j). The problem is to find v points in t space such that $(i_1, j_1) \lesssim (i_2, j_2)$ implies $d_{i_1j_1} \leq d_{i_2j_2}$, that is d as a function on X is isotonic (order-preserving).

Kruskal [16], [17], proposes a measure of goodness of fit to the isotonic relationship by a prescribed set of points. Let $\widehat{d}$ denote the isotonic regression of d on X. The 'raw stress' of the v points is defined as

$$\sum_{j=1}^{v} \sum_{i=1}^{j-1} (d_{i,j} - \widehat{d}_{i,j})^2 .$$

A normalized stress is defined as the square root of the quotient of the raw stress by the sum of squares of the distances. Programming techniques are used to determine the v points so as to minimize the stress. This description of t dimensional scaling is taken from [6]. For $t = 1$, we will get a ranking of the v objects given the dissimilarities ascertained from the paired-comparison design.

Other graph theoretic methods of ranking for incomplete paired comparisons by several judges as a generalization of techniques used in [21], [22] and [14] will be discussed in a later communication.

References

[1] Archbold, J. W. and Johnson, N. L. (1958): A construction for Room's squares and an application in experimental design. Ann. Math. Statist. *29*, 219–225.

[2] Bose, R. C. (1956): Paired-comparison designs for testing concordance between judges. Biometrika *43*, 113–121.

[3] Bose, R. C. and Shimamoto, T. (1952): Classification and analysis of partially balanced incomplete block designs with two associate classes. J. Amer. Statist. Assoc. *47*, 151–184.

[4] Berlekamp, E. R. and Hwang, F. K. (1972): Constructions for balanced Howell rotations for bridge tournaments. Journal of Combinatorial Theory (A) *12*, 159–166.

[5] Bradley, R. A. and Terry, M. E. (1952): Rank analysis of incomplete block designs. I. The method of paired comparisons. Biometrika *39*, 324–345.

[6] Barlow, R. E. et al. (1972): Statistical Inference under Order Restrictions. John Wiley and Sons, New York.

[7] Bühlmann, H. and Huber, P. (1963): Pairwise comparisons and ranking in tournaments. Ann. Math. Statist. *34*, 501–510.

[8] Brunk, H. D. (1960): Mathematical models for ranking from paired comparisons. J. Amer. Statist. Assoc. *55*, 503–520.

[9] Calvin, L. D. (1954): Doubly balanced incomplete block designs for experiments in which the treatment effects are correlated. Biometrics *10*, 61–88.

[10] Chakravarti, I. M. (1961): On some methods of construction of partially balanced arrays. Ann. Math. Statist. *32,* 1181–1185.

[11] David, H. A. (1963): The Method of Paired Comparisons. Griffin, London.

[12] David, H. A. (1971): Ranking the players in a round robin tournament. Review of the Int. Statist. Institute *39,* 137–147.

[13] Huber, P. J. (1963): Pairwise comparison and ranking: optimum properties of the row sum procedure. Ann. Math. Statist. *34,* 511–520.

[14] Kadane, J. B. (1966): Some equivalence classes in paired comparisons. Ann. Math. Statist. *37,* 488–494.

[15] Kendall, M. G. (1955): Further contributions to the theory of paired comparisons. Biometrics *11,* 43–62.

[16] Kruskal, J. B. (1964): Non-metric multidimensional scaling: a numerical method. Psychometrika *29,* 115–129.

[17] Kruskal, J. B. (1964): Multidimensional scaling by optimizing goodness of fit to a non metric hypothesis. Psychometrika *29,* 1–27.

[18] Moon, J. W. (1968): Topics on Tournaments. Holt, Rinehart and Winston, New York.

[19] Schellenberg, P. J. (1973): On balanced Room squares and complete Howell rotations. Aequationaes Mathematicae *9,* 75–90.

[20] Scheffé, H. (1952): An analysis of variance for paired comparisons. J. Amer. Statist. Assoc. *47,* 381–400.

[21] Thompson, W. A., Jr. and Remage, R. (1964): Rankings from paired comparisons. Ann. Math. Statist. *35,* 739–747.

[22] Remage, R., Jr. and Thompson, W. A., Jr. (1966): Maximum likelihood paired comparison rankings. Biometrika *53,* 143–149.

[23] Room, T. G. (1955): A new type of magic square. Math. Gazette *39,* 307.

[24] Wielandt, H. (1950): Unzerlegbare, nicht negative Matrizen. Math. Zeit. *52,* 642–648.

[25] Wilkinson, J. W. (1957): An analysis of paired-comparison designs with incomplete repetitions. Biometrika *44,* 97–113.

[26] Williams, E. G. (1949): Experimental designs balanced for the estimation of residual effects of treatments. Australian J. Scient. Res. *A 2,* 149–168.

[27] Zermelo, E. (1929): Die Berechnung der Turnier-Ergebnisse als ein Maximalproblem der Wahrscheinlichkeitsrechnung. Math. Zeit. *29,* 436–460.

Author's address:
I. M. Chakravarti, Department of Statistics, University of North Carolina at Chapel Hill.

Lucien Féraud

Lucien Féraud

Un modèle d'élimination et de croissance

1. Un modèle simple

Un organisme vivant ou une partie d'un organisme vivant peut soit périr soit subsister et dans ce cas nous supposerons qu'il peut évoluer, par exemple, croître ou plus généralement accroître l'un de ses facteurs caractéristiques (poids, volume, composition chimique, etc.).

Pour rester dans le domaine des variables discrètes, nous supposerons qu'une élimination ne peut se produire qu'au bout d'un nombre entier d'intervalles de temps et nous représenterons par x_i le nombre de ces intervalles de temps vécus, à une certaine époque, par l'organisme: pour abréger nous dirons que x_i est l'«âge» de l'organisme et nous le ferons varier depuis un âge inférieur x_1 jusqu'à un âge supérieur x_n (l'âge pouvant être compté selon un intervalle de temps autre que l'année). Nous supposerons en outre que le facteur caractéristique que l'on considère ne peut que croître, que prendre des valeurs entières

$$s_1, s_2, \ldots, s_j, \ldots, s_m \quad \text{(croissantes)}$$

et que le passage de l'une à celle qui lui est immédiatement supérieurs ne peut avoir lieu qu'à la fin d'un des intervalles de temps ci-dessus introduits.

Nous considérons une population dont la répartition d'après l'âge x_i et le facteur s_j est à chaque époque, définie par une fonction $n(x_i, s_j, t)$ de trois variables que nous supposerons ne prendre que des valeurs entières; en abrégé nous noterons cette fonction $n_{ij}(t)$ ou seulement n_{ij} lorsque l'époque n'aura pas à être rappelée. Le passage d'une époque t à la suivante $t + 1$ pourra entrainer une élimination, par exemple une mortalité, que définira une probabilité de survie de x_i à x_{i+1} supposée indépendante de t, et que nous représenterons par $p(x_i)$ et en abrégé par p_i. En ce qui concerne le facteur s nous supposerons que ne peut se manifester qu'une croissance de s_j à s_{j+1} avec une probabilité λ_{ij} ou l'arrêt à s_j avec une probabilité $1 - \lambda_{ij}$.

En d'autres termes les probabilités de passage, entre t et $t+1$, de l'état i, j à l'état i', j' sont ainsi définies

$$\begin{aligned}
&\lambda_{ij/i'j'} = 0 && \text{si} \quad i' \neq i+1 \,, \\
&\lambda_{ij/i+1,j'} = 0 && \text{si} \quad j' \notin (j, j+1) \,, \\
&\lambda_{ij/i+1,j+1} = \lambda_{ij}\, p_i && \text{avec} \quad \lambda_{im} = 0 \quad \text{et} \quad p_n = 0 \,, \\
&\lambda_{ij/i+1,j} = (1-\lambda_{ij})\, p_i \,.
\end{aligned}$$

Au cours d'un intervalle de temps l'évolution du nombre des individus de la population et leur répartition selon les deux facteurs considérés seront régies par la relation de récurrence

$$n_{i+1,j+1}(t+1) = n_{ij}(t)\, p_i\, \lambda_{ij}(t) + n_{ij+1}(t)\, p_i[1 - \lambda_{ij+1}(t)] \tag{1}$$

avec $\lambda_{im}(t) = 0$, $p_n = 0$,

$$i = 1, 2, \ldots, n \,,$$

$$j = 1, 2, \ldots, m \,.$$

Pour tenir compte des entrées dans la population, s'il en est, nous supposerons – selon la même simplification que précédemment – qu'elles ne se produisent qu'aux instants qui séparent deux intervalles de temps et par suite à un âge x_i et un facteur s_j: nous noterons $e_{ij}(t)$ le nombre des entrées de $t-1$ à t que nous supposerons se produire à l'instant t, à l'âge x_i, au facteur s_j.

Si l'on fait abstraction de la répartition de la population selon le facteur s et que l'on pose[1] $n_{i\cdot}(t) = \sum_j n_{ij}(t)$ et $e_{i\cdot}(t) = \sum_j e_{ij}(t)$, les $n_{i\cdot}(t+1)$, $n_{i\cdot}(t)$ et les $e_{i\cdot}(t)$ sont liés par

$$\left.\begin{aligned}
&n_{1\cdot}(t+1) = e_{1\cdot}(t+1) \\
&n_{i+1\cdot}(t+1) = e_{i+1\cdot}(t+1) + n_{i\cdot}(t)\, p_i
\end{aligned}\right\} \tag{2}$$

pour $i = 1, 2, \ldots, n-1$.

Ces relations définissent la survie de t à $t+1$ des individus de chaque âge. Elles permettent de déterminer

a) la répartition par âge à $t+1$ si l'on donne la répartition par âge à t, les entrées à chaque âge et la mortalité;

[1]) Sauf indication contraire les sommations en j se feront sur $j = 1, 2, \ldots, m$ et les sommations en i se feront sur $i = 1, 2, \ldots, n$.

b) les entrées à chaque âge si l'on se donne la répartition par âge à t et à $t+1$ et la mortalité;

c) la mortalité si l'on donne les répartitions par âge à t et à $t+1$ et les entrées à chaque âge.

Si, au contraire, on fait abstraction de l'âge x et que l'on considère exclusivement la répartition de la population selon le facteur s en posant $n_{.j}(t) = \sum_i n_{ij}(t)$ et $e_{.j}(t) = \sum_i e_{ij}(t)$ on arrive au système de relations

$$n_{.1}(t+1) = e_{.1}(t+1) + \sum_i n_{i1}(t)\, p_i[1 - \lambda_{i1}(t)\,, \tag{3}$$

$$n_{.j}(t+1) = e_{.j}(t+1) + \sum_i n_{ij-1}(t)\, p_i\, \lambda_{ij-1}(t) + \sum_i n_{ij}(t)\, p_i[1 - \lambda_{ij}(t)]\,,$$

pour $j = 2, \ldots, m$.

Ces relations conduisent immédiatement à trois problèmes.

En premier lieu elles donnent la répartition de la population à $t+1$ selon le facteur s en partant de la répartition de la population à t selon les deux variables, des entrées réparties selon le facteur s, des probabilités de survie et des probabilités de croissance.

Un deuxième problème se pose si l'on prend pour inconnues les entrées. Les relations (2) et (3) sont en nombre $n+m$ mais elles ne sont pas indépendantes car la somme totale des entrées tirée des relations (2) coincide avec la somme totale tirée des relations (3): l'une et l'autre représentent la différence entre le total de la population à $t+1$ et le total provenant à la même date des survivants de la population qui existait à t.

Si l'on prend pour inconnues les $e_{i.}$ et les $e_{.j}$, en nombre $n+m-1$ les relations (2) et (3) déterminent une solution unique qui aura une signification lorsqu'elle sera toute entière positive ou nulle, cela à des conditions qui sont apparentes; si l'on prend pour inconnues les e_{ij}, en nombre $n\,m$ il existera, si les mêmes conditions sont satisfaites, une infinité de solutions positives ou nulles parmi lesquelles on pourra, en résolvant un programme linéaire, déterminer celles qui réalisent tel ou tel optimum que l'on décidera d'imposer.

Nous nous arrêterons davantage sur le problème que l'on pose en prenant pour inconnues les λ_{ij} et en les supposant indépendants de t. Ils sont au nombre de $n\,m$ et ne sont liés que par $m-1$ relations linéaires et par suite largement indéterminés. Pour qu'ils définissent des probabilités il faut qu'ils soient compris entre 0 et 1: les conditions s'obtiennent aisément en procédant comme dans le cas particulier qui va être envisagé maintenant.

Il est intéressant de considérer le cas où les λ ne dépendent que du facteur s: nous les noterons λ_j. On peut assimiler à ce cas celui où $\lambda(x, s) = \varphi(x)\,\lambda(s)$ car il suffirait d'agréger $\varphi(x)$ à $p(x)$ et le raisonnement portant sur les $\lambda(s)$ resterait le même.

Le système (3) s'écrit alors, en allégeant les notations et posant $T_j = \sum_i n_{ij} p_i$

$$\left.\begin{aligned} &n_1 = e_1 + (1 - \lambda_1)\, T_1 \\ &\dots \\ &n_j = e_j + \lambda_{j-1}\, T_{j-1} + (1 - \lambda_j)\, T_j \quad \text{pour} \quad j = 1, 2, \dots, m\,, \end{aligned}\right\} \tag{4}$$

et encore

$$\left.\begin{aligned} &n_1 - e_1 = (1 - \lambda_1)\, T_1 \\ &\dots \\ &n_1 + n_2 + \cdots + n_j - e_1 - e_2 - \cdots - e_j - T_1 - T_2 - \cdots \\ &\qquad - T_{j-1} = (1 - \lambda_j)\, T_j \\ &\dots \\ &n_1 + n_2 + \cdots + n_m - e_1 - e_2 \cdots - e_m - T_1 - T_2 \\ &\qquad - \cdots - T_m = 0\,. \end{aligned}\right\} \tag{5}$$

Les données étant telles qu'elles vérifient la dernière de ces relations, les $m - 1$ premières relations déterminent une solution et une seule

$$\text{en} \quad \lambda_1, \lambda_2, \dots, \lambda_{m-1}$$

et mettent en évidence les conditions auxquelles les données doivent satisfaire pour que cette solution définisse des probabilités c'est-à-dire des nombres compris entre 0 et 1.

Les relations ci-dessus s'écrivent encore

$$N_j = E_j + \mathcal{T}_j - \lambda_j T_j = E_j + \mathcal{T}_{j-1} + (1 - \lambda_j)\, T_j$$

en désignant par N_j, E_j, $\mathcal{T}_j$ les cumulatives des n_j, e_j, T_j respectivement.

Il apparait ainsi que la cumulative N_j correspondant à des probabilités λ_j $(0 < \lambda_j < 1)$ est toujours au dessus de la cumulative $E_j + \mathcal{T}_j - T_j$ que l'on obtiendrait avec des λ tous égaux à 1: ce qui reviendrait à supposer qu'à chaque passage d'un intervalle de temps au suivant toute la population s'accroît d'un facteur s_j au suivant (à l'exception de celle qui est déjà en s_m). Par contre la dite cumulative N_j sera toujours au dessous de la cumulative $E_j + \mathcal{T}_j$ que l'on obtiendrait avec des λ_j tous nuls c'est-à-dire en supposant qu'aucun accroissement du facteur s ne se produit. On voit encore que la condition nécessaire et suffisante pour que la cumulative d'une répartition $n(\lambda^{(1)})$

obtenue avec des probabilités $\lambda^{(1)}$ soit toujours au-dessus de la cumulative de la répartition $n(\lambda^{(2)})$ obtenue avec des $\lambda^{(2)}$ (toutes choses égales par ailleurs) est $\lambda_j^{(1)} < \lambda_j^{(2)}$.

Remarquons qu'au lieu de se borner au cas particulier $\lambda(s)$ on pourrait introduire les probabilités d'accroissement moyennes

$$\bar{\lambda}_j = \bar{\lambda}(s_j) = \frac{\sum_i n_{ij}\, p_i\, \lambda_{ij}}{\sum_i n_{ij}\, p_i}$$

et l'on obtiendrait pour les $\bar{\lambda}_j$ les mêmes formules et les mêmes conclusions que pour les λ_j.

2. **Modèles plus généraux**

Le modèle simple considéré jusqu'ici peut se généraliser tout en restant dans le domaine des variables discrètes: on n'astreindra plus la croissance à ne pouvoir se faire que d'un niveau s_j au niveau immédiatement supérieur s_{j+1}, elle sera définie par la probabilité $\lambda_{irk}(t)$ qu'à l'âge x_i l'organisme qui subsiste passe du niveau s_r au niveau s_k; avec

$$r \leqslant k \quad \text{et} \quad \sum_{k=r\dots m} \lambda_{irk}(t) = 1\,, \quad \forall_{i,r,t}\,.$$

Les relations (3) se généralisent

$$n_{.k}(t+1) = e_{.k}(t+1) + \sum_i \sum_{r=1,\dots k} n_{ir}(t)\, p_i\, \lambda_{irk}(t)\,,$$

d'où par sommation en k (de $k = 1$ à $k = j$)

$$N_j(t+1) = E_j(t+1) + \sum_i \sum_{k=1,\dots j} \sum_{r=1,\dots k} n_{ir}(t)\, p_i\, \lambda_{irk}(t)\,,$$

qui s'écrit aussi

$$N_j(t+1) = E_j(t+1) + \sum_i \sum_{r=1,\dots j} \sum_{k=r,\dots j} n_{ir}(t)\, p_i\, \lambda_{irk}(t)$$

et encore à l'aide de

$$\sum_{k=r,\dots j} \lambda_{irk}(t) = 1 - \sum_{k=j+1,\dots m} \lambda_{irk}(t)\,,$$

$$N_j(t+1) = E_j(t+1) + \sum_i \sum_{r=1,\dots j} n_{ir}(t)\, p_i - \sum_i \sum_{r=1,\dots j} \sum_{k=j+1,\dots m} n_{ir}(t)\, p_i\, \lambda_{irk}(t)\,. \quad (6)$$

Le deuxième terme du deuxième membre de cette relation représente le nombre et la répartition en s des organismes qui existaient à t et qui survivent à $t+1$ dans l'hypothèse où il n'y aurait aucun accroissement (tous les λ seraient nuls); en le notant $\mathcal{T}_j(t+1)$ la relation (6) s'écrit

$$N_j(t+1) = E_j(t+1) + \mathcal{T}_j(t+1) - \sum_i \sum_{r=1,\dots j} \sum_{k=j+1,\dots m} n_{ir}(t)\, p_i\, \lambda_{irk}(t)\,. \quad (7)$$

Cette expression se généralise en passant aux variables continues: en représentant par $n(x, s, t)$ l'évolution du nombre des organismes dans l'hypothèse où il n'y aurait pas de croissance (au facteur s constant) et posant

$$\mathcal{T}(s, t) = \int_{x_1}^{x_n} dx \int_{s_1}^{s} n(x, s, t)\, ds$$

on obtient

$$N(s, t) = E(s, t) + \mathcal{T}(s, t) - \int_{x_1}^{x_n} dx \int_{s_1}^{s} dr \int_{s}^{s_m} n(x, r, t)\, \lambda(x, r, k, t)\, dk \quad (8)$$

analogue à (5) et à (7).

3. Courbes de Lorenz et coefficient de Gini

A une époque donnée t les courbes de Lorenz sont données en X et Y par les deux équations

$$X(s) = \frac{1}{N} \int_{s_1}^{s} dN(u) \quad \text{avec} \quad X(s_m) = 1\,,$$

$$Y(s) = \frac{1}{M} \int_{s_1}^{s} u\, dN(u) \quad \text{avec} \quad Y(s_m) = 1\,.$$

$N(u)$ étant la fonction de répartition de la distribution $n(.\,, u, t)$, l'aire sous la courbe de Lorenz par

$$A = \int_0^1 Y\, dX = \frac{1}{M\,N} \int_{s_1}^{s_m} dN(s) \left[\int_{s_1}^{s} u\, dN(u) \right] = 1 - \frac{1}{M\,N} \int_{s_1}^{s_m} s\, N(s)\, dN(s)$$

et le coefficient de Gini par $G = 1 - 2\,A$.

Soit deux populations dont les répartitions selon le facteur s sont définies par les fonctions de répartition

$N_1(s)$	et	$N_2(s)$	qui conduisent à
N_1, M_1	et	N_2, M_2	les totaux correspondants
A_1	et	A_2	les aires sous les courbes de Lorenz
G_1	et	G_2	les coefficients de Gini.

L'inégalité $A_1 > A_2$ équivaut à

$$\frac{1}{M_1 N_1} \int_{s_1}^{s_m} s\, N_1(s)\, dN_1(s) < \frac{1}{M_2 N_2} \int_{s_1}^{s_m} s\, N_2(s)\, dN_2(s)$$

et si $M_1 N_1 = M_2 N_2$ à

$$\int_{s_1}^{s_m} s[N_1(s)\, dN_1(s) - N_2(s)\, dN_2(s)] < 0 \,,$$

$$\int_{s_1}^{s_m} s\, d[N_1^2(s) - N_2^2(s)] < 0$$

et par parties

$$\int_{s_1}^{s_m} [N_1^2(s) - N_2^2(s)]\, ds > 0 \,, \tag{9}$$

si nous supposons $N_1 = N_2$ (et par suite $M_1 = M_2$).

De plus

$$\int_{s_1}^{s_m} s\, d[N_1(s) - N_2(s)] = 0$$

et par parties

$$\int_{s_1}^{s_m} [N_1(s) - N_2(s)]\, ds = 0 \,.$$

Pour que l'inégalité (9) soit vérifiée, il suffit, en tenant compte de la croissance de $N_1(s) + N_2(s)$ que $N_1(s) - N_2(s)$ soit d'abord négative puis positive.

Supposons que les deux populations soient issues d'une même population qui aurait évolué selon $\mathcal{T}(s, t)$ s'il n'y avait pas eu de croissance, qui aurait reçu les entrées $E(s, t)$ mais dont les répartitions en s résulteraient de probabilités de croissance différentes $\lambda_{xrk}^{(1)}(t)$ et $\lambda_{xrk}^{(2)}(t)$. En vertu de (8)

$$N_1(s, t) - N_2(s, t) = - \int_{x_1}^{x_n} n(x, s, t)\, dx \int_{s_1}^{s} dr \int_{s}^{s_m} [\lambda_{xrk}^{(1)}(t) - \lambda_{xrk}^{(2)}(t)]\, dk\,.$$

Dans le cas particulier où les λ ne dépendent pas de x et pour un t déterminé

$$N_1(s) - N_2(s) = - [\Lambda^{(1)}(s) - \Lambda^{(2)}(s)] \int_{x_1}^{x_n} n(x, s)\, dx$$

en posant

$$\Lambda^{(1)}(s) = \int_{s_1}^{s} dr \int_{s}^{s_m} \lambda_{rk}^{(1)}\, dk$$

et une expression analogue pour $\Lambda^{(2)}(s)$.

Il s'ensuit que si la différence $\Lambda^{(1)}(s) - \Lambda^{(2)}(s)$ évolue de telle sorte qu'elle soit d'abord positive puis négative, l'inverse aura lieu pour la différence $N_1(s) - N_2(s)$. Par suite, en admettant $N_1 = N_2$ et $M_1 = M_2$ cette évolution est une condition suffisante pour $A_1 > A_2$ c'est-à-dire $G_1 < G_2$.

En conclusion, pour que le coefficient de Gini augmente – et par conséquent la concentration au sens de ce coefficient – il suffit que les probabilités $\Lambda^{(1)}(s)$ soient plus fortes que les $\Lambda^{(2)}(s)$ pour les basses valeurs de s et plus faibles pour les hautes valeurs, ceci pour une même population totale et pour une même valeur globale du facteur s. L'interprétation de $\Lambda(s)$ est évidente: pour chaque valeur de s c'est la probabilité de passage par dessus la valeur s, d'une valeur inférieure à une valeur supérieure.

Adresse de l'auteur:
Lucien Féraud, 1, rue Viollier, 1207 Genève.

Léopold J. Martin

Léopold J. Martin

Intérêt du diagramme (β_1, β_2) de K. Pearson et du diagramme (γ_1, γ_2) dans le champ bio-médical

Résumé

Présentation d'un diagramme $(\beta_1; \beta_2)$ de K. Pearson dans le domaine $(0 < \beta_1 < 6; 1 < \beta_2 < 9)$, ainsi que du diagramme $(\gamma_1; \gamma_2)$ correspondant.

1. Diagramme (β_1, β_2) de K. Pearson

En général, le diagramme de K. Pearson est présenté dans le rectangle limité par $0 < \beta_1 < 1{,}8$ et $1 < \beta_2 < 8$ (voir par exemple E. S. Pearson et Hartley, 1954, p. 210). K. Pearson (1933) a cependant présenté une esquisse couvrant un domaine plus large $(0 < \beta_1 < 6{,}0; 0 < \beta_2 < 14{,}8)$.

Nous avons présenté un diagramme détaillé couvrant le domaine $0 < \beta_1 < 4; 1 < \beta_2 < 9)$ et qui objective des convergences intéressantes au point G (0,3) et au point E (4,9). (Martin [3], p. 36, voir Fig. 1.)

A noter: trois aires particulières importantes: l'aire impossible (A. I), l'aire hétérotopique (A. H) pour le système de K. Pearson et l'aire occupée par les lois bêta.

1.1 *Points (β_1, β_2) et segments disposés sur l'axe $\beta_1 = 0$, donc répondant à des densités symétriques*

β_2	Segments		
1,0		B_1	Binomiale positive $(p + q)^n$ avec $n = 1$
1,5		F	Bêta (1/2, 1/2), densité arc sinus ou $\sin^{-1}$ Feller (1958 et 1966)
	$B \rightarrow R$		Type II (U) de K. Pearson (antimode .5)

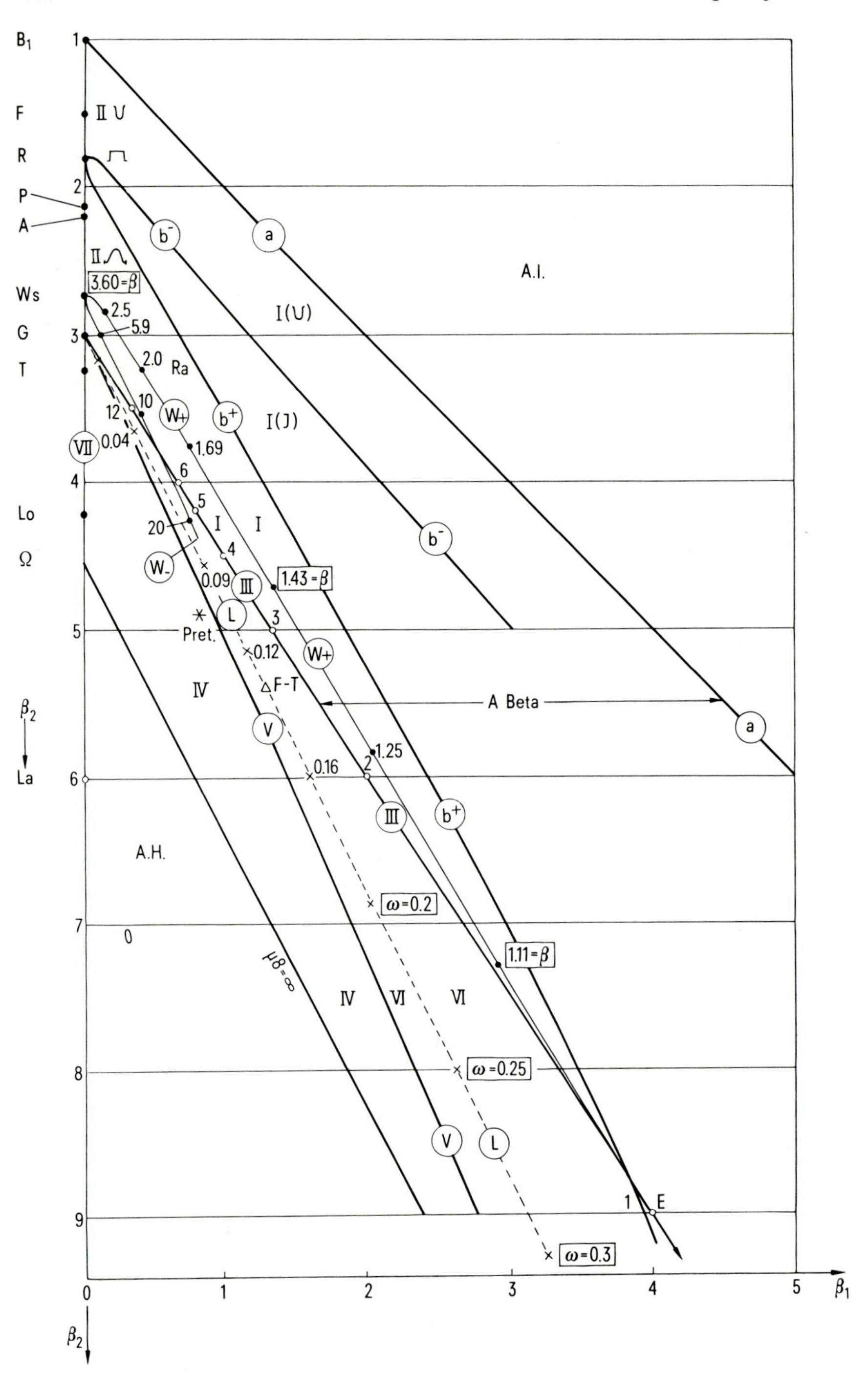

B1
F
R
P
A
Ws
G
T
Lo
Ω
La
A.I.
A.H.
A Beta
I(∪)
I(J)
3.60=β
1.43=β
1.11=β
ω=0.2
ω=0.25
ω=0.3
Pret.
F-T
Ra
E
μ8=∞
β2
β1

Fig. 1

1.1 (Suite)

β_2	Segments		
1,8		R	Rectangulaire ou Bêta (1, 1)
2,14		P	Parabolique ou Bêta (2, 2)
2,2		A	Angulaire symétrique
2,66		W_0	Weibull pseudo-symétrique ($\mu_3 = 0$) au paramètre $\beta \approx 3{,}60$
	$R \to G$		Type II en cloche de K. Pearson
3,0		G	Gauss; point de convergence des lignes III, L et V
3,25		T	Triangulaire symétrique
4,2		Lo	Logistique, Gumbel (1958, p. 311)
	$G \to \Omega$		Type VII de K. Pearson
4,5		Ω	Origine de la droite-limite ($\mu_3 = \infty$) pour le système de K. Pearson
6,0		La	Laplace (Martin [4], p. 372)

1.2 *Lignes, Zones et Points en dehors de l'axe* $\beta_1 = 0$

(a) Droite limite pour toutes les lois de probabilité.

(III) Droite lieu des points des courbes de *type Pearson III*, l'un des trois types principaux du système de K. Pearson.

Ces courbes sont aussi appelées lois gamma — l, au paramètre l. Les points sont cotés en l.

$l = 1$: *Point E*; *densité exponentielle négative* qui répond aussi au type X de K. Pearson. Coordonnées: $\beta_1 = 4$, $\beta_2 = 9$.

A partir du point E, la flèche indique la direction dans laquelle se trouve le point ($\beta_1 = 8$; $\beta_2 = 15$) correspondant à la *densité* χ^2 *avec* 1 *degré de liberté*.

A noter également qu'au point E convergent deux lignes importantes:

a) la ligne ($W+$) répondant aux lois de Weibull à asymétrie positive (paramètre $\beta < 3{,}60$),

b) la ligne (III), lieu des densités gamma au paramètre l.

Si $l \to \infty$, les points gamma — l tendent vers le point G, *densité gaussienne* de coordonnées $\beta_1 = 0$, $\beta_2 = 3$.

L'aire bêta ou zone des lois Pearson I est divisée en trois régions par la courbe (b) limitant la zone $I(J)$.

La courbe supérieure notée (b —) répond aux lois Pearson I à $\mu_3 < 0$.

Le point R est un point de transition ($\mu_3 = 0$) vers la courbe inférieure notée ($b+$) qui répond aux lois Pearson I à $\mu_3 > 0$.

Lignes ($W+$) *et* (W —), lieu des points (β_1, β_2) répondant à la famille des lois de Weibull d'exposant β qui apparaît comme cote sur ces lignes.

Distribution de Weibull réduite:

$$F(y) = 1 - \exp(-y^{\beta})$$

où β est un paramètre de forme (shape parameter).

Application microbiologique et virologique (voir Martin [4], pp. 523–524).

Cas particuliers:

$\beta = 1$ *Point E.*

Exponentielle négative ou loi du *type Pearson X* ($\beta_1 = 4$; $\beta_2 = 9$).

$\beta = 2$ *Point Ra.*

Loi de Rayleigh (*Ra*). Martin (1961, 1962 et 1973, p. 510). Coordonnées ($\beta_1 \approx .40$; $\beta_2 \approx 3{,}25$).

Application cardiologique ainsi qu'aux courbes de survie.

$\beta \approx 3{,}60$ *Point* W_0.

Loi de Weibull pseudosymétrique.

W_0 = point de transition entre la ligne ($W+$) et la ligne ($W-$) et répondant à la condition $\beta_1 = 0$.

La ligne ($W+$) partant du point E reste cantonnée dans la zone de type I et est toujours située «au-dessus» de la ligne (III) des lois gamma.

La ligne ($W-$) part de W_0 pour $\beta = 3{,}60$ et traverse la droite (III) en un point situé en $\beta_1 \approx .7$ et $\beta_2 \approx 3{,}75$.

La loi de Weibull au paramètre $\beta = 10$ est encore située dans la zone du type K. Pearson I. Enfin pour $\beta = 20$, les coordonnées du point représentatif sont $\beta_1 = .75$ et $\beta_2 = 4{,}25$; ce point tombe dans la zone du type VI de K. Pearson. Voir aussi E. S. Pearson (1969).

Ligne (V).

Lieu des points répondant au type V de K. Pearson.

Zone du type VI située entre les lignes (III) et (V), tant dans la zone non hétérotopique que dans la zone hétérotopique (vide infra n° 33). Le type VI est le deuxième des trois types principaux de K. Pearson.

Lieu (L) *des points des densités log-gaussiennes* (log-normales) au paramètre ω

$$\omega = \gamma^2 = (\sigma/\mu)^2.$$

Utilisées, entre autres, par Boag (1948) dans le follow-up de patients traités par radiothérapie. Voir aussi Aitchison et Brown (1957) et Martin [4].

Pour les densités log-gaussiennes habituelles, β_1 est positif. Dans le graphique 1, les points (x) sont cotés en $\omega = 0{,}30$; 0,20, etc., jusqu'à 0,01. Ce dernier point est proche du point G répondant à la loi de Gauss.

On remarque que la ligne (L) est située dans la zone répondant au type VI de K. Pearson. La ligne (L), notée aussi S_L par Johnson (1949, p. 157) est la ligne centrale du *système de N. L. Johnson* (1949). Elle sépare d'une part la *famille* S_B (B pour bounded, c'est-à-dire marge finie) qui occupe l'aire comprise entre les lignes (a) et (L). D'autre part, la *famille* S_U (U pour unbounded) occupe l'aire située «sous» la ligne (L).

Point F–T, situé pratiquement sur la ligne (L).

Fisher et Tipett (1928) ont introduit la première distribution de l'extrême supérieur en la variable réduite y.

$$F(y) = \exp(-e^{-y}) \quad -\infty < y < \infty .$$

Gumbel (1958, 1960, pp. 157–162) a largement développé cette loi pour laquelle $\beta_1 = 1{,}3$ et $\beta_2 = 2{,}4$.

Droite-limite ($\mu_8 = \infty$) pour le système des lois de K. Pearson.

Cette droite débute au point Ω pour lequel $\beta_2 = 4{,}5$.

Zone du type Pearson IV située entre la ligne (V) et la droite-limite qui vient d'être considérée. Les distributions du type Pearson IV appartiennent au deuxième des types principaux de K. Pearson.

Le point * Pret. de coordonnées $\beta_1 = .83$ et $\beta_2 = 4{,}86$ répondant à la distribution des longueurs de 9440 haricots (beans) est indiqué dans la zone IV du diagramme 1. Sur ces données se sont escrimés Pretorius (1930), K. Pearson (1933), Johnson (1949), Draper (1952), Kendall et Stuart (1958) et Martin (1958).

Dans la graduation des tableaux de fréquences, les courbes du type Gram-Charlier ou Edgeworth jouent un rôle important.

Dans le diagramme (β_1, β_2), les zones à densités unimodales d'une part et à fréquences attendues positives d'autre part sont données d'après Barton et Dennis dans Elderton et Johnson (1969, p. 121).

2. **Diagramme** (γ_1, γ_2)

Dans une communication au Congrès de l'Institut International de Statistique (Vienne, 20–30 août 1973), nous avons présenté la version définie dans le rectangle ($-3 < \gamma_1 < +3$; $-2 < \gamma_2 < +6$) (Martin [5], Fig. 2). avec est consvire $\gamma_2 = \beta_2 - 3$. La ligne (a) est dédoublée en les arcs de parabole (a —) et (a +). Pour la famille des lois bêta aux paramètres l et m

$$f(x) = K\, x^{l-1}\, (1-x)^{m-1} \quad (0 < x < 1)$$

les deux courbes (b —) et (b +) de la Fig. 2 répondent à la limite $I(J)$ du diagramme (β_1, β_2) et au lieu des points (γ_1, γ_2) pour $l = 1$. Les courbes (b' —) et (b' +) leur sont symétriques et correspondent à $m = 1$.

Dans les six zones ainsi définies, sont localisées les courbes bêta avec anti-, pseudo- et vrai mode et possédant une asymétrie soit positive, soit négative.

A la classique droite du type III dans le diagramme (β_1, β_2), répondent deux arcs de parabole (III —) et (III +).

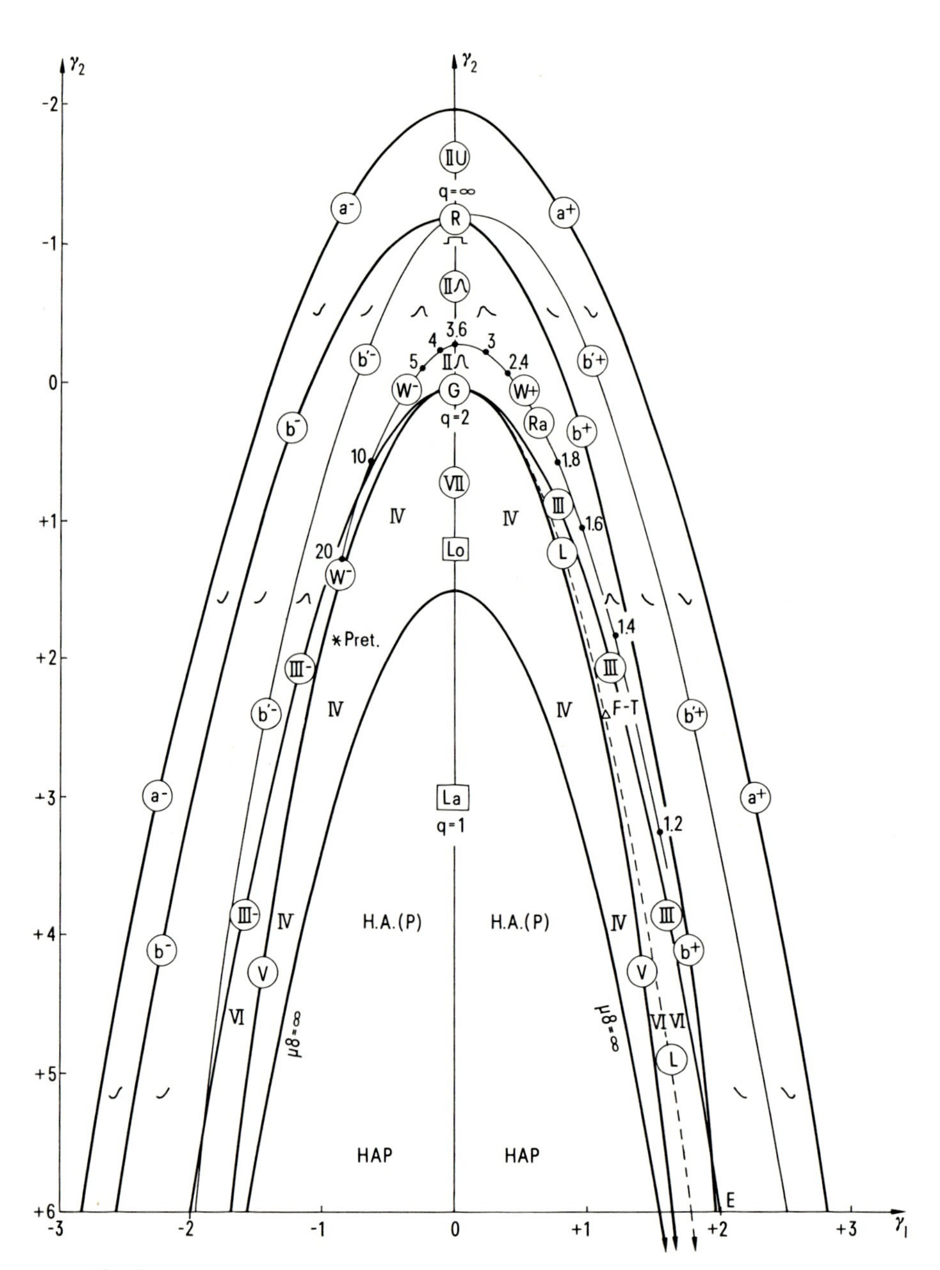

γ_2
γ_1
ⅡU
q=∞
R
ⅡΛ
3.6
4
3
5
2.4
W⁻
G
W+
q=2
Ra
10
1.8
Ⅶ
Ⅲ
Ⅳ
1.6
Lo
L
20
W⁻
1.4
*Pret.
Ⅲ-
F-T
La
q=1
1.2
H.A.(P)
HAP
V
Ⅵ
μ8=∞
E
a⁻
a+
b′⁻
b′+
b⁻
b+

Fig. 2

Les courbes $(W -)$ et $(W +)$ répondent à la famille des lois de Weibull d'asymétrie négative et positive avec transition en la loi de Weibull pseudo-symétrique définie par $\beta \approx 3{,}60$.

La ligne (L) en pointillé dans le diagramme (γ_1, γ_2) est le lieu des lois log-gaussiennes d'asymétrie positive. La ligne V répond à la ligne V du diagramme (β_1, β_2); la parabole $(\mu_8 = \infty)$ répond à la droite correspondante en (β_1, β_2).

Sur l'axe $\gamma_1 = 0$, les cotes $q = 1$ (Laplace), $q = 2$ (Gauss) et $q = \infty$ (Rectangulaire) indiquent trois cas de lois symétriques considérées par Box [1] dans un contexte plus général.

Points particuliers: point F–T dans la région $\gamma_1 > 0$; point Pretorius dans la région $\gamma_1 < 0$.

Dans une étude récente, Johnson et Kotz [2] ont reporté dans un diagramme $\sqrt{\beta_1}$ (ou γ_1) et $\beta_2 = 3 + \gamma_2$ les points représentatifs des lois suivantes:

a) lois en transformées puissance (X^λ) et logarithmique d'une variable gamma (Pearson III);
b) la densité puissance $\theta\, z^{\theta-1}$ $(0 < z < 1; \theta > 0)$;
c) la densité puissance inverse $\theta\, z'^{-(\theta+1)}$ $(z' > 1; \theta > 0)$.

Nous renvoyons le lecteur à cette note très intéressante où en particulier une base rationnelle est donnée à la transformation bien connue du χ^2 selon Wilson et Wilferty $(\lambda = 1/3)$.

Bibliographie

Pour épargner de l'espace typographique, nous renvoyons pour la bibliographie à Martin (1972/73). Les références complémentaires sont les suivantes:

[1] Box, G. E. P. (1953): A note on regions for tests of kurtosis. Biometrika *40*, 465–468.

[2] Johnson, N. L. and Kotz, S. (1972): Power transformations of gamma variables. Biometrika *59*, 226–229.

[3] Martin, L. (1972/73): Le diagramme (β_1, β_2) de K. Pearson dans le domaine biomédical. Biométrie-Praximétrie *13*, 33–44.

[4] Martin, L. (1973 a): Distributions and Transformations. Chap. 6 of Biostatistics in Pharmacology, 155–610 (Sect. 7, vol. I of International Encyclopaedia of Pharmacology and Therapeutics), Pergamon Press, Oxford.

[5] Martin, L. (1973 b): Graphs and diagrams for continuous distributions including a (γ_1, γ_2) diagram. Présenté à la 39e session de l'Institut International de Statistique (Vienne, 20–30, août 1973).

Adresse de l'auteur:
Léopold J. Martin, Professeur de Statistique Médicale, Université libre de Bruxelles, Faculté de Médecine et Ecole de Santé Publique, 100, rue Belliard, B–1040 Bruxelles.

J. N. Srivastava

Some Further Theory of Search Linear Models

J. N. Srivastava

Summary

Let $\boldsymbol{y}(N \times 1)$ be a vector of observations, $A_1(N \times v_1)$ and $A_2(N \times v_2)$ known matrices, $\xi_1(v_1 \times 1)$ and $\xi_2(v_2 \times 1)$ vectors of parameters and σ^2 a constant such that

$$E(\boldsymbol{y}) = A_1 \xi_1 + A_2 \xi_2 , \tag{1}$$

$$V(\boldsymbol{y}) = \sigma^2 I_N . \tag{2}$$

Here σ^2 may be known or unknown, $\xi_1(v_1 \times 1)$ is unknown. About ξ_2, we have partial information. It is known that there is a positive integer k, such that at most k elements of ξ_2 are non-zero, the rest being negligible. However, it is not known which k elements of ξ_2 are (possibly) non-zero. The problem is to search the non-zero elements of ξ_2, and make inferences about these, and about the elements of ξ_1. Such models, called 'Search Models', were introduced in Srivastava [1]. In this paper, some further basic developments are made in this direction.

1. Introduction

Consider the above search model. The case $\sigma^2 = 0$, called 'the noiseless case', plays a fundamental role. Take the following estimation problem: We wish to search the non-zero elements of ξ_2 and estimate them and also the elements of ξ_1. It was shown in Srivastava [1] that, in the noiseless case, a necessary and sufficient condition that this search-cum-estimation problem can be 'completely solved' it is that for *every* $(N \times 2\,k)$ submatrix A_2^* of A_2, we must have

$$\text{Rank} \quad (A_1 \mid A_2^*) = v_1 + 2\,k . \tag{3}$$

By 'completely solved', we mean that we will be able to search the non-zero set of elements of ξ_2 without any error, and furthermore obtain estimators which have variance zero. A set of observations $\boldsymbol{y}$, for which Condition (3) is satisfied, are said to form a 'search design'.

In practical statistical problems, we shall always have $\sigma^2 > 0$. However, even here the noiseless case is important, since any difficulties arising in the noiseless case are inherently present when σ^2 is assumed positive. For example, the Condition (3) still remains necessary for the above search-cum-estimation problem to be solved completely. However, obviously, it is no longer sufficient. Note that sufficiency does not hold even when no search is involved.

In this paper, we develop some further basic theory under the noiseless case.

2. The Case $v_1 > 0$

In many important applications, we have $v_1 > 0$. For example, if ξ_1 denotes the set consisting of the general mean μ and the main effects in an $s_1 \times \ldots \times s_m$ factorial experiment, then for a design of resolution III (or 'main effect plan'), we have

$$v_1 = 1 + \sum_{i=1}^{m} (s_i - 1) , \quad \text{and} \quad v_2 = (\prod_{i=1}^{m} s_i) - \sum_{i=1}^{m} s_i + (m - 1) ,$$

where the v_2 effects consisting of the 2-factor and higher order interactions are assumed zero. However, as is well known, there has been almost no situation (where experimenters normally use a main effect plan), where all of the said v_2 effects would actually be negligible. On the other hand, it is also simultaneously true that in these same situations, most of the v_2 effects are indeed negligible. Thus the true situation is this. There is a positive integer k_0, where k_0 is very very small relative to v_2, such that out of the above v_2 effects, at most k_0 effects are non-zero. In other words, the search model (1, 2) is applicable with

$$v_1 = \sum_{i=1}^{m} s_i - m + 1 , \quad v_2 = (\prod_{i=1}^{m} s_i) - v_1 , \quad k = k_0 . \tag{4}$$

Note that in earlier work with experimenters using main effect plans, the bias caused by the unknown set of k_0 non-zero effects did vitiate the results. The above search theoretic formulation thus constitutes a realistic and complete solution to the problem. Of course, as is true of all statistical problems, the amount of success finally achieved in correctly carrying out the search and estimation will depend upon the size of the noise present.

The above remarks also apply to other situations. For example, in the above, we could have talked of a resolution V design in place of resolution III. The values of the parameters would then be

$$v_1 = 1 + \sum_{i=1}^{m} (s_i - 1) + \sum_{\substack{i>j \\ i,j=1,\ldots,m}} [(s_i - 1)(s_j - 1)] ,$$

$$v_2 = (\prod_{i=1}^{m} s_i) - v_1 , \quad k = k_0 . \tag{5}$$

Now, Condition (3) says that we must take sets of $(v_1 + 2\,k_0)$ columns of the matrix $[A_1 \mid A_2]$ including v_1 columns of A_1, and check if in each case the $N \times (v_1 + 2\,k_0)$ submatrix so formed has full rank. Since v_1 itself may be large in many cases, for example in (4) and (5) above, the $N \times (v_1 + 2\,k)$ submatrices to be checked will be relatively quite large. This shows the importance of the following result which throws the problem back to the case when $v_1 = 0$.

Theorem 1. Consider the model (1, 2), where $\mathbf{y}$ etc., are partitioned as below

$$\mathbf{y} = \left[\frac{\mathbf{y}_1}{\mathbf{y}_2}\right], \quad A_1 = \left[\frac{A_{11}}{A_{21}}\right], \quad A_2 = \left[\frac{A_{12}}{A_{22}}\right], \tag{6}$$

where $\mathbf{y}_1$, A_{11} and A_{12} have n_1 rows each, and $N = n_1 + n_2$. Then a sufficient condition that (3) holds for every A_2^* is that (7) below holds:

(i) $A_{11}(n_1 \times v_1)$ has rank v_1, and (7)

(ii) there exist matrices $B_1(l \times n_1)$ and $B_2(l \times n_2)$, where l is an arbitrary positive integer, such that

$$B_2\,A_{21} = B_1\,A_{11}\,, \tag{8}$$

and the matrix $(B_2\,A_{22} - B_1\,A_{12})$ has property P_{2k}. (From Srivastava [1], recall that a matrix M is said to have property P_t if every set of t columns of M are linearly independent.)

Proof: Let $A_2^*(N \times 2\,k)$ be a submatrix of A_2 such that $[A_1 \mid A_2^*]$ has rank less than $(v_1 + 2\,k)$. Let the partitioning $A_2' = [A_{12}' \mid A_{22}']$ induce the partitioning $A_2^{*\prime} = [A_{12}^{*\prime} \mid A_{22}^{*\prime}]$. There exists a non;null vector $\mathbf{g}\big((v_1 + 2\,k) \times 1\big)$ such that $[A_1 \mid A_2^*]\ \mathbf{g} = 0(n \times 1)$. Partition $\mathbf{g}$ as $\mathbf{g}' = [\mathbf{g}_1' \mid \mathbf{g}_2']$, where $\mathbf{g}_1$ is $(v_1 \times 1)$. Thus we get $A_1\,\mathbf{g}_1 + A_2^*\,\mathbf{g}_2 = 0$. This implies

$$\mathbf{O} = [B_1 \mid -B_2]\,[A_1\,\mathbf{g}_1 + A_2^*\,\mathbf{g}_2] = [B_1 \mid -B_2]\left\{\left[\frac{A_{11}}{A_{21}}\right]\mathbf{g}_1 + \left[\frac{A_{12}^*}{A_{22}}\right]\mathbf{g}_2\right\}$$

$$= (B_1\,A_{11} - B_2\,A_{21})\,\mathbf{g}_1 + (B_1\,A_{12}^* - B_2\,A_{22}^*)\,\mathbf{g}_2 = (B_1\,A_{12}^* - B_2\,A_{22}^*)\,\mathbf{g}_2\,,$$

in view of (8). Now the $(l \times v_2)$ matrix $(B_1\,A_{12} - B_2\,A_{22})$ possesses the property P_{2k}, and contains the $(l \times 2\,k)$ submatrix $(B_1\,A_{12}^* - B_2\,A_{22}^*)$. Hence we must have $\mathbf{g}_2 = \mathbf{0}$. Thus $A_1\,\mathbf{g}_1 = \mathbf{0}$, implying that the columns of A_1 are dependent. But this contradicts (7.1). This completes the proof.

Theorem 2: In Theorem 1, the choice

$$l = n_2\,, \quad B_2 = I_{n_2}\,, \quad B_1 = A_{21}(A_{11}'\,A_{11})^{-1}\,A_{11}'\,, \tag{9}$$

satisfying (8) is always possible.

Proof: Obvious.

Theorem 3: Under the model (1, 2), the search and estimation problem concerning ξ_2 can be separated from the estimation problem for ξ_1.

Proof: Using (6), let

$$\boldsymbol{u} = B_2\, \boldsymbol{y}_2 - B_1\, \boldsymbol{y}_1\,, \tag{10}$$

where B_2 and B_1 are as at (9). Then

$$E(\boldsymbol{u}) = (B_2\, A_{22} - B_1\, A_{12})\, \xi_2 = \big(A_{22} - A_{21}(A'_{11}\, A_{11})^{-1}\, A'_{11}\, A_{12}\big)\, \xi_2\,,$$
$$= Q\, \xi_2\,, \tag{11}$$

say, and

$$\mathrm{Var}(\boldsymbol{u}) = \sigma^2[I_{n_2} + A_{21}(A'_{11}\, A_{11})^{-1}\, A_{21}] = \sigma^2(K\, K')\,, \tag{12}$$

say, where $K(n_2 \times n_2)$ is non-singular. Then $\boldsymbol{u}^* = K^{-1}\, \boldsymbol{u}$ satisfies (1, 2) with $v_1 = 0$, and A_2 replaced by $K^{-1}\, Q$. Now, clearly, $K^{-1}\, Q$ has property P_{2k} if and only if Q does. Hence if Q has the property P_{2k}, then using $\boldsymbol{u}^*$, one can first slove the search-estimation problem for ξ_2. Having done this, one could substitute in (1), and solve the estimation problem for ξ_1 separately. We now show that if (3) holds, then there always exists a choice of A_{11} and hence B_1 such that the resulting Q will have P_{2k}. To see this, first notice that A_1 must have rank v_1, and hence A_1 must contain a submatrix A_{11} with $n_1 = v_1$ and rank v_1. Consider B_1, B_2 and Q under this choice of A_{11}. Now,

$$\left[\begin{array}{c|c} -A_{21}(A'_{11}\, A_{11})^{-1}\, A'_{11} & I_{n_2} \\ \hline I_{n_1} & 0 \end{array}\right] \left[\begin{array}{c|c} A_{11} & A_{12} \\ \hline A_{21} & A_{22} \end{array}\right] = \left[\begin{array}{c|c} 0 & Q \\ \hline A_{11} & A_{12} \end{array}\right]. \tag{13}$$

The first matrix on the left hand side of (13) is clearly non-singular. Hence if $[A_1 \mid A_2]$ satisfies (3), then so does the matrix on the right hand side of (13). Now since A_{11} is a square non-singular matrix, it clearly follows that Q must have P_{2k}. This completes the proof.

Remark 1. The last theorem would be very useful both (i) from the point of view of constructing search designs, and (ii) for a given $\boldsymbol{y}$, conducting the search and estimation. The reason is that v_1 is usually large compared to k which would usually be quite small. The search is really concerned with only k effects, and so the above separation of ξ_1 and ξ_2 will be helpful.

3. 'Resolution V plus' Plans, and 'Main Effect Plus' Plans

We partly discussed these earlier, at Equations (4) and (5). Consider (5) for 2^m factorial. Then $v_1 = 1 + m(m+1)/2$, we need to obtain search designs, i.e. sets T of assemblies, such that using T, Condition (3) be satisfied. The matrix $[A_1 \mid A_2]$ will be of size $(N \times 2^m)$ with rows corresponding to assemblies in T, and columns corresponding to the effects in ξ_1 and ξ_2; the element of this matrix corresponding to the assembly $(t_1, \ldots, t_m)$ and effect $F_1^{j_1} \ldots F_m^{j_m}$ will be $\alpha(\pi_{i=1}^m \varepsilon_i)$, where α is a constant independent of the assembly and the effect, and $\varepsilon_i = -1$, if $j_i = 1$ and $t_i = 0$, and $\varepsilon_i = 1$ otherwise. Notice that the column in $[A_1 \mid A_2]$ corresponding to the effect $F_1^{j_1} \ldots F_m^{j_m}$ is the product of the columns corresponding to the main effect $F_r^{j_r}$ (in which $j_r \neq 0$). Thus the matrix $[A_1 \mid A_2]$ is highly structured. Needless to say that this structure should be exploited for studies on property P_{2k} etc.

Now suppose that $v_1 > 0$. If we use Theorems 1–3, we need to have $\boldsymbol{y}_1$ and $\boldsymbol{y}_2$ such that $Q^* = A_{22} - A_{21}(A_{11}' A_{11})^{-1} A_{11}' A_{21}$ has property P_{2k}. Suppose $\boldsymbol{y}_1$ corresponds to a set of treatments T_1 and $\boldsymbol{y}_2$ to T_2. The question is what property should T_1 and T_2 have so that Q^* may have P_{2k}. Now, as we indicated above, $[A_1 \mid A_2]$ possesses a rich structure. But what about Q^*? We make a few informal remarks here regarding these questions on the basis of our experience so far, for the benefit of potential researchers. Formal work will require many series of papers.

For arbitrary but fixed T_1, Q^* usually has no easily recognisable pattern which could be directly connected with T_2. A method of selection of T_1 which at present seems to provide the maximum ability for relating T_2 and Q^* is as follows. We choose T_1 so that it is invariant under a renaming of the factors. Equivalently, if we write T_1 as an $(n_1 \times m)$ matrix whose rows represent treatment-combinations, and columns represent factors, then T should be invariant under a permutation of columns. Another equivalent condition is that T_1 be a balanced array of strength m.

Now for 2^m factorials, optimal balanced resolution V plans have been obtained by Srivastava and Chopra in a series of papers, for $m = 4, 5, 6, 7, 8$, and for a practical range of values of n_1 (number of assemblies) for each m. Fortunately, it so turns out that each of these optimal designs is a B-array of full strength (i.e. m). Thus it might be good to conduct work on the following lines. What is a 'good' set of treatments T_2 to be added to an optimal B-array T_1, so that we may obtain a 'resolution V plus' plan.

A beginning has been made in the study of 'resolution III plus' plans for 2^m factorials by Srivastava and Gupta. Here T_1 consists of the $(m+1)$ treatments $(1, 1, \ldots, 1)$, $(1, 0, 0, \ldots, 0)$, $(0, 1, 0, \ldots, 0)$, $\ldots$, $(0, 0, \ldots, 0, 1)$. The optimal balanced plans of resolutions III usually turn out to be balanced arrays of strength 2 only, and give rise to additional difficulties when used as T_1.

4. Weakly Resolvable Search Models

We now introduce another very important concept in the theory of search models. Firstly, as a simple example, take the special case of (1, 2) with $\sigma^2 = 0$, $v_1 = 0$, $N = 3$, $k = 1$, $v_2 = 3$, given by

$$y_1 = \xi_1 + \xi_2 + 2\,\xi_3\,, \quad y_2 = 3\,\xi_1 - \xi_2 + 6\,\xi_3\,,$$

$$y_3 = 3\,\xi_1 - \xi_2 - 2\,\xi_3\,. \tag{14}$$

If only (y_1, y_2) are taken, then (3) is not satisfied, since clearly ξ_1 and ξ_3 are confounded. Similar situation holds if we take (y_1, y_3), or (y_2, y_3). Thus at first sight it appears that to solve the search-estimation problem, we need all the three observations y_1, y_2, and y_3. While this is true if all observations are taken in one stage, it is not necessarily true if sequential experimentation is considered. Thus, suppose at the first stage, we use (y_1, y_2). If ξ_2 happens to be non-zero, then y_1, y_2 will be in the ratio 1 to (-1), and vice versa. Hence, in the first stage, if $y_1 = -y_2$, the search-estimation problem gets resolved. Same thing happens if $y_1 = y_2 = 0$. The problem is unresolved only when $y_2 = 3y_1$, in which case we will not known whether ξ_1 or ξ_3 is non-zero, and a further observation (e.g. y_3) is needed.

A search model of the type (1, 2) is said to be *strongly resolvable,* if (under the noiseless case) we can completely solve the search-estimation problem. Thus Condition (3) is necessary and sufficient for strong resolvability. A search model which is not strongly resolvable, is said to be *weakly resolvable.*

In the above weakly resolvable model with (y_1, y_2), the parameters ξ_1 and ξ_3 will be separable for certain values of the observations and certain kinds of a priori information on the unknown parameters. For example, if it is known a priori that $\xi_1 \leq 2$, and we find that $y_2 = 3y_1 = 8$, then clearly we must have $\xi_1 = \xi_2 = 0$, $\xi_3 = (4/3)$. Notice that the information on ξ_1 is only in the form of an inequality rather than an exact value.

We now present an example to show that sometimes when even a meagre amount of a prioir information is available, it can be used to make an otherwise weakly resolvable model strongly resolvable. Thus, suppose, under (1, 2), for $\sigma^2 = 0$, $v_1 = 0$, $v_2 = 4 = N$, $k = 2$, we have

$$\begin{bmatrix} y_1 \\ y_2 \\ y_3 \\ y_4 \end{bmatrix} = \begin{bmatrix} 1 & 3 & 2 & -2 \\ 2 & -5 & -3 & 0 \\ 3 & 1 & 0 & -4 \\ 0 & 2 & 3 & 1 \end{bmatrix} \begin{bmatrix} \xi_1 \\ \xi_2 \\ \xi_3 \\ \xi_4 \end{bmatrix}, \text{ or } y = B\,\xi\,. \tag{15}$$

Here the vector $(1, 1, -1, 1)$ is orthogonal to the rows of B (which has rank 3), so that B does not have property P_4. However suppose it is known that ξ_1 does *not* equal any one among ξ_2, $(-\xi_3)$, or ξ_4. Notice that, relatively speaking, this is only a small amount of information. We leave it to the reader to check that with this extra information, (15) becomes strongly resolvable.

In general, if S is a set of statements concerning ξ_1, ξ_2, such that under the knowledge of S, an otherwise weakly resolvable model M becomes strongly resolvable, then M is said to be S-resolvable.

Theorem 4: Consider the search model (1, 2), and suppose it is weakly resolvable. Let $(i_1, \ldots, i_k)$ be a set of distinct integers (with $i_1 < i_2 < \cdots < i_k$) out of the set $L = \{1, 2, \ldots, v_2\}$, and let K be the collection of all the $\binom{v}{k_2}$ such sets. Let $(i_{k+1}, \ldots, i_{k+p})$, $1 \leq p \leq k$, be integers (with $i_{k+1} < \cdots < i_{k+p}$) from L, such that $i_r\,(r=1,\ldots,k+p)$ are all distinct. Let $A_2^{**} = A_2^*(i_1,\ldots,i_{k+p})$ be the $n \times (k+p)$ sub-matrix of A_2 formed by the columns number $i_1, \ldots, i_{k+p}$ from A_2. Suppose that the columns of A_1 together with any set of $(k+p-1)$ columns of A_2^{**} are linearly independent. Let $V_p(A_1, A_2^{**})$ be the vector space orthogonal to the rows of $(A_1 \mid A_2^{**})$. Let $W = W(i_1, \ldots, i_k)$ be the collection of all vectors $(w_1, \ldots, w_k)$ such that there exist $\alpha_j\,(j=1,\ldots,v_1)$ and $\beta_j\,(j = 1, \ldots, p)$ such that for some p, and for some vector $(i_{k+1},\ldots,i_{k+p})$, the vector $(\alpha_1, \ldots, \alpha_{v_1}, w_1, \ldots, w_k, \beta_1, \ldots, \beta_p)$ belongs to $V_p(A_1, A_2^{**})$. Suppose S is the collection of statements that for all $(i_1, \ldots, i_k) \in K$, there does not exist a set of parameters $(\xi_{i_1}, \ldots, \xi_{i_k})$ with value $(w_1^*, \ldots, w_k^*)$, such that the vectors $(w_1^*, \ldots, w_k^*)$ and $(w_1, \ldots, w_k)$ have identical element in p or more places $(1 \leq p \leq k)$, and $(w_1, \ldots, w_k) \in W$. Then the model is S^*-resolvable if and only if S^* contains S.

Proof: Suppose $(\xi_{i_1}, \ldots, \xi_{i_k})$ is the non-zero set and has value $(w_1^*, \ldots, w_k^*)$. Suppose the model is S^*-resolvable and S^* contains S, Then $(w_1^*, \ldots, w_k^*) \in W$. Suppose $\xi_1' = (\alpha_1^*, \ldots, \alpha_{v_1}^*)$, and suppose $\boldsymbol{a}_r$ and $\boldsymbol{b}_r$ denote the r-th columns of A_1 and A_2 respectively. Then (in the noiseless case), we have

$$\boldsymbol{y} = \sum_{r=1} \alpha_r^* \,\boldsymbol{a}_r + \sum_{r=i_1}^{i_k} w_r^* \,\boldsymbol{b}_r \,.$$

This projection of $\boldsymbol{y}$ on the column space of $[A_1 \mid A_2]$ must be unique. If not, suppose we have another projection

$$\boldsymbol{y} = \sum_{r=1} \boldsymbol{a}_r^{**} \,\boldsymbol{a}_r + \sum_{r=j_1}^{j_k} w_r^{**} \,\boldsymbol{b}_r$$

tnvolving columns $(j_1, \ldots, j_k)$ of A_2. Suppose, for simplicity, that between the sets $(i_1, \ldots, i_k)$, $(j_1, \ldots, j_k)$, there are exactly p elements common, say, $i_1, \ldots, i_p$. Notice that we must have $p \geq 1$, since otherwise A_1 will be of rank iess than v_1. Then the vector $\big((\alpha_1^* - \alpha_1^{**}), \ldots, (\alpha_{v_1}^* - \alpha_{v_1}^{**}), (w_1^* - w_1^{**}), \ldots, (w_p^* - w_p^{**}), w_{p+1}^*, \ldots, w_k^{**}, - w_{p+1}^{**}, \ldots, - w_k^{**}\big)$ belongs to $V_{k-p}(A_1, A_2^+)$, where A_2^+ has columns of A_2 corresponding to the distinct integers in $(i_1, \ldots, i_k, j_1, \ldots, j_k)$.

Thus $\big((w_1^* - w_1^{**}), \ldots, (w_p^* - w_p^{**}), w_{p+1}^*, \ldots, w_k^*\big) \in W$, and has $(k - p)$ elements common with $(w_1^*, \ldots, w_k^*)$, a contradiction. Hence if S^* contains S than the model is S^*-resolvable. Clearly, the converse can be easily established by reversing the above argument. This completes the proof.

The above result is fundamental in the theory of weakly resolvable models. These models are obviously of considerable potential importance both from the point of view of sequential and one-stage experimentation.

Search models, though formulated only a short while ago, are finding increasing applications in various disciplines. This includes for example single factor experiments like varietal trials. The idea here is to utilize a priori information regarding the structure of the varieties by bringing in a search approach. This however will be considered elsewhere because of lack of space.

References

[1] SRIVASTAVA, J. N. (1973): Designs for searching Non-negligable Effects. To appear in *A Survey of Statistical Design and Linear Models*. North Holland Publishing Company, Amsterdam.

[2] (For other relevant references see the above paper.)

Author's address:
J. N. Srivastava, Colorado State University and Indian Statistical Institute.

W. J. Ziegler

W. J. Ziegler

Zum Problem der Optimum-Eigenschaften von SPR-Tests

Zur Entwicklung des SPR-Tests

Nachdem in der früheren Testtheorie, begründet von J. Neyman und E. S. Pearson, der Stichprobenumfang konstant geblieben war, gestaltete ihn A. Wald variabel, wobei er für den Verlauf des so geprägten sequentiellen Tests als Regel drei mögliche Entscheidungen herleitet: Nach jedem Beobachtungsschritt ist die einfache Hypothese $H_0(\theta = \theta_0)$ gegen die einfache Alternative $H_1(\theta = \theta_1)$ anzunehmen, abzulehnen, oder es wird eine weitere Beobachtung gefordert.

Wald bezeichnet $f(x, \theta)$ als Verteilung der Zufallsvariablen x, die im kontinuierlichen Fall eine Wahrscheinlichkeitsdichtefunktion annimmt oder diskret verteilt ist und bezeichnet $x_1, \ldots, x_m$ als Stichprobenraum R_m von x. Er gibt für jeden positiven Integralwert m die Wahrscheinlichkeit an, dass man eine Stichprobe $x_1, \ldots, x_m$ erhält mit

$$p_{1m} = f(x_1, \theta_1) \cdot \ldots \cdot f(x_m, \theta_1)\,, \quad \text{wenn } H_1 \text{ richtig ist}$$

und

$$p_{0m} = f(x_1, \theta_0) \cdot \ldots \cdot f(x_m, \theta_0)\,, \quad \text{wenn } H_0 \text{ richtig ist.}$$

Wohl nicht exakt, aber willkührlich kann man für das Prüfen einfacher Hypothesen zwei Konstanten A_0' und A_1' (wobei $A_0' \neq A_1'$) wählen, und es gibt bei jedem Stand der Beobachtungsreihe und damit bei der m-ten Bestimmung des Integralwertes m das **S**equential **P**robability **R**atio zu p_{1m}/p_{0m}.

Die SPR's sind logarithmisch als Summen darstellbar und der Testprozess bewegt sich im indifferenten Bereich

$$A_0 < \sum_{i=1}^{m} \log \frac{f(x_i, \theta_1)}{f(x_i, \theta_0)} = \sum_{i=1}^{m} z_i < A_1\,,$$

bis das Testkriterium $\Sigma\, z_i$ eine der Stoppregeln erfüllt.

Im Vergleich mit den traditionellen Testkonstruktionen mit fixen Stichprobenumfängen haben die Aussagen über die Erwartungswerte für die Stichprobenumfänge überrascht, und es liegt nahe, die Optimum-Eigenschaft nicht nur für den SPR-Test mit einzelnen Beobachtungen, sondern für solche mit ein oder wenigstens zwei Beobachtungen pro Testschritt zu untersuchen.

Optimum-Eigenschaft des SPR-Tests

Das Theorem über die Optimum-Eigenschaft besagt, dass der SPR-Test unter allen Stichprobenumfängen m, für die

$$p_0 \ \text{(Ablehnung von } H_0) \leq \alpha_0 \text{ und } p_1 \text{ (Ablehnung von } H_1) \leq \alpha_1$$

gilt, mit den Fehlerwahrscheinlichkeiten α_0 und α_1, die Erwartungswerte $E_0(m)$ und $E_1(m)$ minimisiert.

Dieses *Theorem* legt eine Erweiterung nahe, die besagt, *dass unter allen SPR-Tests mit konkaver Kostenfunktion der Beobachtungsfolge der Test mit optimaler Gruppierung der m Beobachtungen in n Versuchen* ($n \leq m$) *am schnellsten zum Testentscheid führt.*

Der Beweis hierfür soll mit einer Besprechung des Falles angegangen werden, die Beobachtungen einzeln oder in Zweiergruppen zu machen:

Für den Test der Hypothese H_0, dass $\theta_i = \theta_0$ die richtige Wahrscheinlichkeitsdichte von x ist, gegen die Alternativhypothese H_1 mit $\theta_i = \theta_1$ seien die Verluste durch irrtümliche Ablehnung von H_0 bzw. H_1 durch ω_0 und ω_1 bezeichnet. Ausserdem seien die experimentellen Kosten einer Beobachtung im Versuch mit c, die Kosten zweier Beobachtungen mit d und der Bedingung anzunehmen, dass $d < 2c$ Ausdruck einer konkaven Kostenfunktion ist. Der zu erwartende Verlust und die Kosten zusammen bilden sodann die Risiken

$$\alpha_i \cdot \omega_i + c \cdot E_i \text{ (Zahl der Versuche mit einer Beobachtung)}$$

und

$$\alpha_i \cdot \omega_i + d \cdot E_i \text{ (Zahl der Versuche mit zwei Beobachtungen)},$$

wenn θ_i $(i = 0, 1)$ die richtige Wahrscheinlichkeitsdichte ist.

Mit dem Subskript i der Wahrscheinlichkeitsdichte als Zufallsvariable lassen sich analog dem Vorgehen von Lehmann mit Testvorgängen $\delta(1)$ mit einer Beobachtung und $\delta(2)$ mit zwei Beobachtungen pro Versuch und den Bayes-Lösungen der Minimalrisiken

$$r(\pi, \delta(1)) = \pi \cdot (\alpha_0 \omega_0 + c \cdot E_0(m)) + (1 - \pi) \cdot (\alpha_1 \omega_1 + c \cdot E_1(m)) \quad (1\text{a})$$

$$r(\pi, \delta(2)) = \pi \cdot \left(\alpha_0 \omega_0 + d \cdot E_0\left(\frac{m}{2}\right)\right) + (1 - \pi) \cdot \left(\alpha_1 \omega_1 + d \cdot E_1\left(\frac{m}{2}\right)\right) \quad (1\text{b})$$

Werte $\pi'(1) \leq \pi'(2) \leq \pi''(2) \leq \pi''(1)$ einschliessen, die eindeutig bestimmt sind durch $\omega_0(1)$, $\omega_0(2)$ und $\omega_1(1)$, $\omega_1(2)$ sowie c bzw. d und die von π unabhängig sind.

Wir folgen der Annahme, dass es genügt, die Situation

$$0 < \pi'(1) \leq \pi'(2) \leq \pi''(2) \leq \pi''(1) < 1 \text{ zu betrachten} \tag{2}$$

mit $\pi'(1) < \pi'(2) < \pi < \pi''(2) < \pi''(1)$ und damit folgendes Lemma:

Lemma:

Wenn $\pi'(1)$, $\pi''(1)$ und $\pi'(2)$, $\pi''(2)$ den Gleichungen (3) genügen und wenn (2) gilt, dann werden die Bayes-Risiken (1) minimisiert

a) für alle $\pi'(1) < \pi < \pi''(1)$
durch jeden SPR-Test mit den Grenzen

$$A_0(1) = \frac{\pi}{1-\pi} \cdot \frac{1-\pi''(1)}{\pi''(1)} \quad \text{und} \quad A_1(1) = \frac{\pi}{1-\pi} \cdot \frac{1-\pi'(1)}{\pi'(1)},$$

ausser:

b) für alle $\pi'(2) \leq \pi \leq \pi''(2)$
durch jeden SPR-Test mit den Grenzen

$$A_0(2) = \frac{\pi}{1-\pi} \cdot \frac{1-\pi''(2)}{\pi''(2)} \quad \text{und} \quad A_1(2) = \frac{\pi}{1-\pi} \cdot \frac{1-\pi'(2)}{\pi'(2)}.$$

Zum Beweis:

Wir beginnen mit der Abschätzung, ob im Versuch wenigstens zwei Beobachtungen gemacht werden sollten mit den Kosten d, ob eine Beobachtung mit den Kosten c oder ob wir besser ohne Beobachtung entscheiden:

δ_0 sei Testvorgang mit Ablehnung von H_0 ohne Beobachtung,
δ_1 sei Testvorgang mit Annahme von H_0 ohne Beobachtung, so dass

$$r(\pi, \delta_0) = \pi \cdot \omega_0 \quad \text{und} \quad r(\pi, \delta_1) = (1-\pi) \cdot \omega_1 .$$

Es seien $\varrho(\pi) = \inf_{\delta \in \mathscr{C}} (\pi, \delta)$

$\mathscr{C}$ ist Klasse aller Versuche mit einer Beobachtung.

$$\tau(\pi) = \inf_{\delta \in \mathscr{D}} (\pi, \delta) .$$

$\mathscr{D}$ ist Klasse aller Versuche mit zwei Beobachtungen.

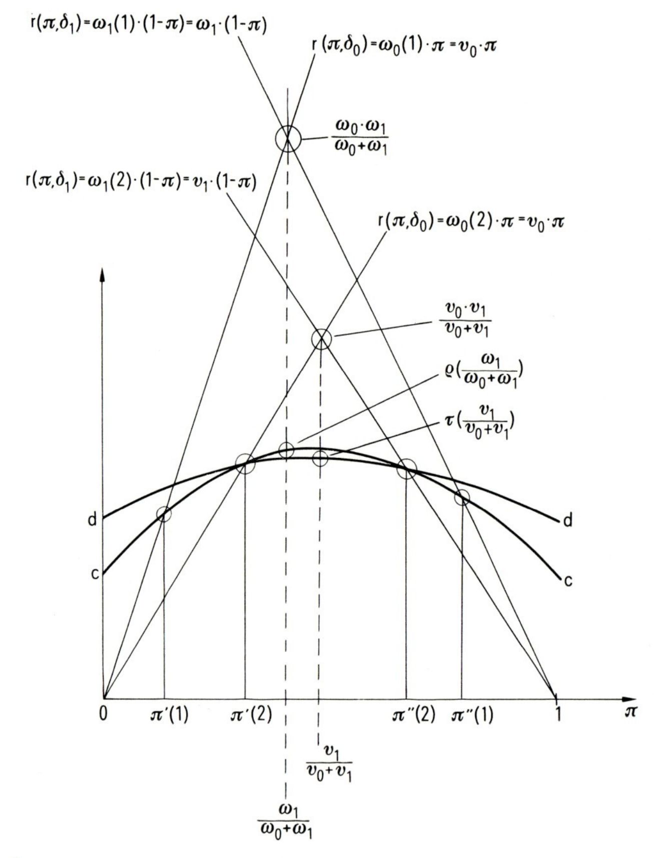

Fig. 1
Illustration der Risikofunktionen bei SPR-Tests mit ein oder zwei Beobachtungen in den Versuchseinheiten.

Für ein jedes $0 < \lambda < 1$ und beliebige π_0, π_1 im Intervall 0, 1

$$\varrho\big(\lambda \cdot \pi_0 + (1-\lambda) \cdot \pi_1\big)$$

$$= \inf_{\delta \in \mathscr{C}}\big(\lambda \cdot r(\pi_0, \delta) + (1-\lambda) \cdot r(\pi_1, \delta)\big) \geq \lambda \cdot \varrho(\pi_0) + (1-\lambda) \cdot \varrho(\pi_1)$$

und für ein jedes $0 < \mu < 1$ und gleiche π_0, π_1

$$\tau\big(\mu \cdot \pi_0 + (1-\mu) \cdot \pi_1\big)$$

$$= \inf_{\delta \in \mathscr{C}}\big(\mu \cdot r(\pi_0, \delta) + (1-\mu) \cdot r(\pi_1, \delta)\big) \geq \mu \cdot \tau(\pi_0) + (1-\mu) \cdot \tau(\pi_1) .$$

ϱ und τ seien konkav und im Intervall 0, 1 kontinuierlich. Für den Fall, dass sich $\varrho(\pi)$ und $\tau(\pi)$ innerhalb $\pi'(1)$ und $\pi''(1)$ schneiden, seien $\pi'(1)$, $\pi'(2)$ und $\pi''(1)$, $\pi''(2)$ mit

$$\tau\left(\frac{v_1}{v_0 + v_1}\right) < \varrho\left(\frac{\omega_1}{\omega_0 + \omega_1}\right) < \frac{v_0\, v_1}{v_0 + v_1} < \frac{\omega_0\, \omega_1}{\omega_0 + \omega_1}$$

definiert durch

$$r(\pi'(1), \delta_0) = \varrho(\pi'(1)) \quad \text{und} \quad r(\pi'(2), \delta_0) > \varrho(\pi'(2)) = \tau(\pi'(2)), \tag{3a}$$

$$r(\pi''(1), \delta_1) = \varrho(\pi''(1)) \quad \text{und} \quad r(\pi''(2), \delta_1) < \varrho(\pi''(2)) = \tau(\pi''(2)). \tag{3b}$$

Ansonsten sei

$$\pi'(2) = \frac{\omega_1}{\omega_0 + \omega_1} = \pi''(2) .$$

Im Fall $0 < \pi'(1) < \pi'(2) < \pi''(2) < \pi''(1) < 1$ minimisieren

δ_0 die Gleichung $r(\pi, \delta)$, wenn $\pi \leq \pi'$,

δ_1 die Gleichung $r(\pi, \delta)$ nur, wenn $\pi \geq \pi''$.

Durch Beobachtung der $x_1, \ldots, x_m$ ändert sich die Wahrscheinlichkeit, dass H wahr ist, a posteriori bei m bereits erfolgten Beobachtungen wie folgt (einseitig betrachtet):

$$\pi(x_1, \ldots, x_m) = \frac{\pi' \cdot p_{0m}}{\pi' \cdot p_{0m} + (1 - \pi') \cdot p_{1m}} .$$

Umgeformt bedeutet dies für das Probability Ratio nach m Beobachtungen schliesslich, dass

$$\frac{p_{1m}}{p_{0m}} = \frac{1 - \pi(x_1, \ldots, x_m)}{\pi(x_1, \ldots, x_m)} \cdot \frac{\pi'}{1 - \pi'} .$$

Das heisst, ein Versuch wird fortgesetzt, mit zwei Beobachtungen je Versuch, solange

$$A_0(2) = \frac{\pi(x_1, \ldots, m)}{1 - \pi(x_1, \ldots, m)} \cdot \frac{1 - \pi''(2)}{\pi''(2)} < \frac{p_{1m}}{p_{0m}} < \frac{\pi(x_1, \ldots, x_m)}{1 - \pi(x_1, \ldots, x_m)}$$

$$\times \frac{1 - \pi'(2)}{\pi'(2)} = A_1(2)$$

mit einer Beobachtung je Versuch, wenn

$$A_0(1) = \frac{\pi(x_1, \ldots, x_m)}{1 - \pi(x_1, \ldots, x_m)} \cdot \frac{1 - \pi''(1)}{\pi''(1)} < \frac{p_{1m}}{p_{0m}} < A_0(2)$$

beziehungsweise wenn

$$A_1(2) = \frac{p_{1m}}{p_{0m}} < \frac{\pi(x_1, \ldots, x_m)}{1 - \pi(x_1, \ldots, x_m)} \cdot \frac{1 - \pi'(1)}{\pi'(1)} = A_1(1) .$$

Der erste Schritt des Testvorganges lässt sich also mit δ_0 für $\pi < \pi'(1)$ und mit δ_1 für $\pi > \pi''(1)$ bestimmen, ferner mit wenigstens einer Beobachtung im Versuch, wenn $\pi'(1) < \pi < \pi'(2)$ bzw. $\pi''(2) < \pi < \pi''(1)$ und mit wenigstens zwei Beobachtungen im weiteren Versuch, wenn $\pi'(2) < \pi < \pi''(2)$.

Wenn $\pi = \pi'(1)$, dann minimisiert der Vorgang δ_0 noch $r(\pi(1), \delta)$, jedoch nicht mehr eindeutig, das heisst, es existiert auch ein Prozess $\delta \in \mathscr{C}$, für den $r(\pi', \delta) = \varrho(\pi')$ und wenn $\pi = \pi'(2)$, dann minimisiert δ_0 noch $r(\pi(2), \delta)$, jedoch nicht mehr eindeutig, das heisst, es existiert auch ein Prozess $\delta \in \mathscr{D}$, für den $r(\pi', \delta) = \tau(\pi')$.

Diskussion

Die aufgezeigten Ansätze sind Erweiterungen gegebener Ableitungen von Lehmann. Sie mögen einerseits dazu stimulieren, die totale Verallgemeinerung zu erreichen. Für konkrete Tests bleiben die Konstruktionen der vier Geraden eines sequentiellen Testprozesses, hierfür die näherungsweise Bestimmung von $\varrho(\pi)$ und $\tau(\pi)$ und damit die näherungsweise Bestimmung der Testkonstanten.

Literatur

[1] LEHMANN E. L. Testing Statistical Hypotheses, Wiley New York 1959.

Adresse des Autors:
W. J. Ziegler, CH-4126 Bettingen/Basel (Schweiz)